Canadian Special Publication of Fisheries and Aquatic Sciences 109

Lingcod

(Ophiodon elongatus)

Alan J. Cass, Richard J. Beamish, and Gordon A. McFarlane

Department of Fisheries and Oceans
Biological Sciences Branch
Pacific Biological Station
Nanaimo, British Columbia V9R 5K6

Scientific Excellence
Resource Protection & Conservation
Benefits for Canadians

DEPARTMENT OF FISHERIES AND OCEANS
Ottawa 1990

Cover photograph: Nest-guarding male lingcod protecting egg mass (Courtesy of Rick Harbo, Department of Fisheries and Oceans, Fisheries Branch, Nanaimo, B.C.)

Published by	Publié par
Fisheries and Oceans	Pêches et Océans
Communications Directorate	Direction générale des communications

Ottawa K1A 0E6

Available from authorized bookstore agents, other bookstores
or you may send your prepaid order to the
Canadian Government Publishing Centre
Supply and Services Canada, Ottawa, Ont. K1A 0S9

Make cheques or money orders payable in Canadian funds
to the Receiver General for Canada

A deposit copy of this publication is also available
for reference in public libraries across Canada.

Catalogue No. Fs 41-31/109E
ISBN 0-660-13429-2
ISSN 0706-6481

DFO/4303

Correct citation for this publication:

CASS, A. J., R. J. BEAMISH, AND G. A. MCFARLANE. 1990. Lingcod (*Ophiodon elongatus*). Can. Spec. Publ. Fish. Aquat. Sci. 109: 40 p.

Contents

Abstract

Cass, A. J., R. J. Beamish, and G. A. McFarlane. 1990. Lingcod (*Ophiodon elongatus*). Can. Spec. Publ. Fish. Aquat. Sci. 109: 40 p.

This publication is a review of the biology and fishery of lingcod. We describe the distribution, reproduction, early life history, diet, age, growth rate and survival rate of lingcod. We also describe the basis for management strategies and present assessments of the condition of the major stocks in Canadian waters.

Lingcod have a long history of exploitation by early native peoples of western Canada. Commercial fishing by white settlers began in the mid-1800s. Before the development of the west coast bottom trawl fishery in the 1940s, lingcod were the main source of fresh fish throughout the year. At that time, lingcod ranked fourth in commercial importance behind salmon, herring and pilchards. Lingcod were also one of the few species of fishes being actively studied by fisheries scientists during the 1920s and 1930s. Lingcod are now an important species in the sports fishery.

This report has been written for scientists, students and the general public. It is meant to provide people interested in marine life with an appreciation of the historical and biological significance of lingcod.

Résumé

Cass, A. J., R. J. Beamish, and G. A. McFarlane. 1990. Lingcod (*Ophiodon elongatus*). Can. Spec. Publ. Fish. Aquat. Sci. 109: 40 p.

Dans le présent survol de la biologie et de l'exploitation de la morue-lingue, les auteurs décrivent la répartition, la reproduction, la genèse, le régime alimentaire, l'âge, le taux de croissance et le taux de survie de l'espèce. Ils élaborent les fondements des stratégies de gestion et des évaluations actuelles de l'État des principaux stocks fréquentant les eaux canadiennes.

De temps immémorial, la morue-lingue a été exploitée par les autochtones de l'Ouest canadien. Le milieu des années 1800 en vit le début de la pêche commerciale par les premiers colons. Avant la mise sur pied de la pêche au chalut de fond sur la côte ouest au cours des années 1940, la morue-lingue représentait la principale source de poisson frais pendant toute l'année et elle se classait au quatrième rang d'importance commerciale après le saumon, le hareng et la sardine du Pacifique. Au cours des années 1920 et 1930, la morue-lingue comptait parmi les quelques espèces qui étaient activement étudiées par des scientifiques. Elle fait maintenant l'objet d'une pêche sportive importante.

Le présent rapport est destiné aux scientifiques, aux étudiants et au grand public. Aux personnes intéressées à la vie marine, il veut leur permettre d'apprécier l'importance historique et biologique de la morue-lingue.

Introduction

Lingcod (*Ophiodon elongatus*) are a unique feature of Canada's west coast fishery. Their large size, aggressive appearance and tasty flesh have attracted fishermen for centuries. In recent years, fishing for lingcod has become so popular that the survival of some stocks is threatened. Despite the popularity of lingcod, it has only been in the last decade that biologists have understood the life history from egg to adult. Effective conservation and management of any species begins with an understanding of this general biology.

We wrote this report so that people interested in marine life can appreciate the complexity of the life cycle of lingcod and to promote the concept that conservation is really the responsibility of all Canadians. As you read through this report, we hope you are impressed with the features of the life history that lingcod have developed to survive in their particular environment. Remember that these features that have evolved slowly over centuries allow lingcod to compete very successfully with other fishes but make it vulnerable to human predation. There is not enough time for nature to select for and develop a population that is wary of humans, thus, if we want our grandchildren and their grandchildren to experience the thrill of observing large lingcod, we need to begin immediately to ensure that adequate numbers of lingcod survive until they are able to reproduce. Hopefully, the information in this report will lead to a better understanding of the biology and management of this magnificent fish and ensure that lingcod remain an important species in the fishery and culture of British Columbians.

The lingcod is not a cod. When early settlers arrived on the West Coast they added the name "cod" to many fishes including the lingcod. It belongs to a little known group of fish called Hexagrammids and is unique to the west coast of North America. The lingcod is one of the largest fish to occur off British Columbia and is a well adapted predator that can grow to 50 or more pounds and lengths of more than 4 feet.

The fishery for lingcod off Canada's west coast probably began between 3 500–5 000 years ago. Lingcod bones have been identified at several early fishing sites along the British Columbia coast that date back thousands of years. A number of Indian groups, including the Coast Salish, Nootka, Kwakiutl, Haida, Coast Tsimshian and Tlingit, used a variety of methods to catch lingcod. For example, the early Coast Salish people developed an ingenious lure made of wood and sinews or roots. This floating lure was pushed below the surface with a long stick. When released, the lure's erratic motion as it rose to the surface attracted the lingcod to the spear or net of a waiting fisherman (Fig. 1).

Lingcod were fished by early settlers because of the large size, tasty flesh and accessibility. Fishing in the mid-1800s occurred mainly in inshore waters around Victoria. As other areas were settled, local fisheries began along eastern Vancouver Island. The methods used to catch lingcod in today's commercial hook and line fishery have changed very little from those used in the early fishery (Fig. 2). Traditionally, lingcod caught in the Strait of Georgia were kept alive in submerged crates until needed for sale (Fig. 3, 4). In recent years the recreational catch has become very important. In fact, lingcod may be more important today as a sport fish in the Strait of Georgia than the current commercial fishery. A large lingcod at the end of a fisherman's line is a thrill that is not soon forgotten. Scuba divers also hunt for lingcod, although in recent years many find pleasure in simply observing these magnificent and colourful fish in their natural environment.

This report has been written for scientists, students and the general public. The method of referencing statements common to scientific literature has not been used. Instead, a list of source material is appended at the end of the report. A glossary is also provided because in some cases it is difficult to avoid the use of scientific terminology. We begin by describing the past and present fisheries and follow with a description of the general biology. The basis for present management strategies is described, followed by a general assessment of the condition of stocks. We conclude with a discussion of management strategies and our opinion of the future of lingcod fisheries.

FIG. 1. Historical method of spearing lingcod by native fishermen.

FIG. 2. Typical handline vessel used in today's commercial lingcod fishery in the Strait of Georgia.

FIG. 3. Handline vessels and submerged crates used to catch and hold commercially caught lingcod in the Strait of Georgia.

FIG. 4. Commercial fishermen processing a catch of lingcod from the Strait of Georgia.

The Fishery

Commercial fishing for lingcod in British Columbia started around 1860. Lingcod were jigged from small vessels less than 10 m long in the sheltered waters of the Strait of Georgia (Fig. 5). Commercial fishing even occurred in Vancouver Harbour during the early 1900s.

The first catch statistics from the commercial fishery were produced in 1889, but the catch of lingcod was not separated from rockfish catches until 1927. It is known that lingcod were the principal species caught during this period, accounting for more than 90% of the catch of "codlike" species. Catches increased gradually between 1889 and 1909 and averaged about 300 tonnes per year (t/yr). As more settlers arrived on Canada's west coast, the markets for fresh lingcod increased. Between 1910 and 1929, catches increased dramatically to an average of nearly 2 500 t. Catches increased during the 1930s and 1940s to meet the high demand for foodfish and liver extracts (vitamin A). During this period catches averaged 3 100 t/yr and reached a maximum of 4 600 t/yr in 1944. By present-day standards this represents a major fishery. Some of the largest catches ever recorded occurred at this time (Appendix Table 1). The good fishing and strong markets of that era attracted many fishermen, unfortunately it is unknown how many vessels were involved. Our only estimates come from recollections of a few fishermen. Mr. Toivo Lane fished lingcod from 1944 to 1985. He recalled that about 200 handline vessels fished lingcod each year from Powell River and Comox to Vancouver and Nanaimo during the "heyday" of the 1940s. Three generations of another family, the Matsunagas, fished each spring along with about 70 other vessels in waters near Cape Mudge until the 1960s. During this time, an estimated 90% of all lingcod caught came from the Strait of Georgia handline fishery. Catches declined after World War II and stabilized at an average of 2 600 t/yr after 1950.

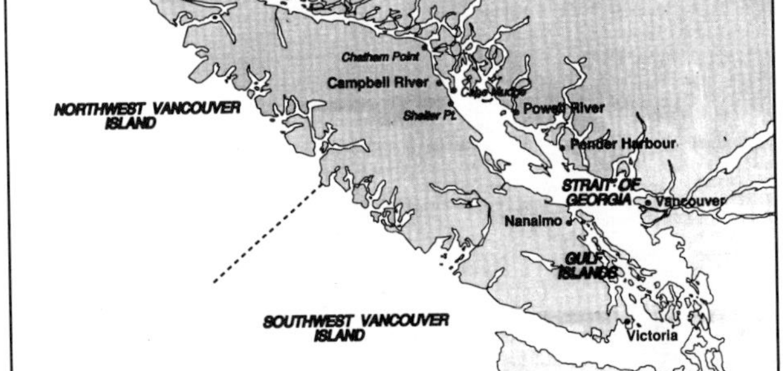

FIG. 5. Lingcod fishing areas off Canada's west coast.

Lingcod were an extremely important source of fresh fish between 1900 and the 1940s. Before the development of the bottom trawl fishery, lingcod were the main source of fresh fish available throughout the year. Compared to other sources of foodfish prior to the 1940s, lingcod ranked fourth in commercial importance behind salmon, herring and the now economically extinct pilchard. Lingcod was also one of the few species of fishes being actively studied by fisheries scientists during the 1920s and 1930s. Only with the expansion of bottom trawling did the emphasis in biological research shift to other species of bottom fish.

The rapidly developing bottom net fishery or trawl fishery that began in the 1940s gradually landed a greater share of the catch of lingcod (Fig. 6). The net fishery also extended the range of exploitation outside the Strait of Georgia. Production from this fishery now dominates the commercial fisheries in all areas except the Strait of Georgia (Fig. 7). Unlike the commercial handline fisheries, which historically targeted on lingcod, bottom trawl-caught lingcod are taken in a mixed species fishery that includes Pacific cod, flatfishes, and rockfishes (Fig. 8).

The Canadian net catch rose steadily from 630 t in 1945 to an average of 800 t/yr in the 1960s. During the 1960s and 1970s, catches fluctuated at an average of 1 400

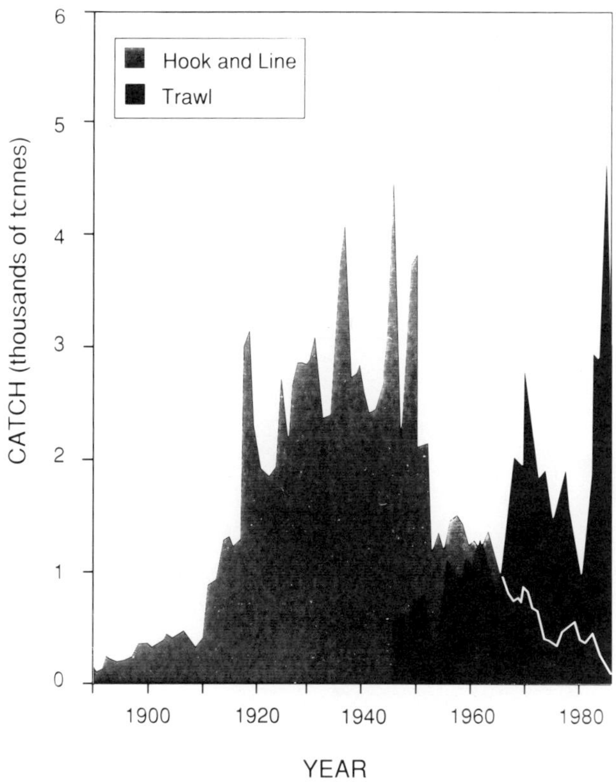

FIG. 6. Canadian catches of lingcod from the commercial hook and line and trawl fisheries off the west coast of Canada.

t/yr with a maximum of 2 900 t in 1968. Between 1980 and 1985, catches increased from 2 100 t to about 5 000 t. In 1985, they were the highest ever recorded.

Historically, both Canadian and United States fishermen participated in the bottom trawl fisheries. The United States fishery was prohibited in most areas in 1978 with the extension of Canada's jurisdiction over offshore resources to 200 miles. For the 1956–78 period of the United States fishery, catches closely paralleled those of the Canadian trawl fleet, averaging 1 300 t/yr (Appendix Table 1).

The main trawling grounds for lingcod have been off southwest Vancouver Island and in Queen Charlotte Sound. Catches from these areas accounted for about 80% of the total Canadian catches during 1956–80. Catches by Canadian and United States fishing fleets in these areas were similar (Fig. 9).

The decline of catches from the handline fishery was a direct result of the decline of catches from the Strait of Georgia. Handline catches in the Strait of Georgia declined from 4 000 t/yr in the mid-1940s to an average of 1 400 t/yr in the 1950s. Handline catches averaged 935 t/yr in the 1960s and 460 t/yr in the 1970s. Between 1980 and 1985, handline catches from the Strait of Georgia declined to an average of 277 t/yr. This represents an 80 percent decline from the catches in the 1950s. The downward trend in catches has continued to the present day. It is interesting that concern for stocks of lingcod in the Strait of Georgia handline fisheries was voiced long before the dramatic decline in catch. As early as 1913, Federal Fisheries Inspectors believed that the intensive fisheries in the Strait of Georgia were overfishing the stocks.

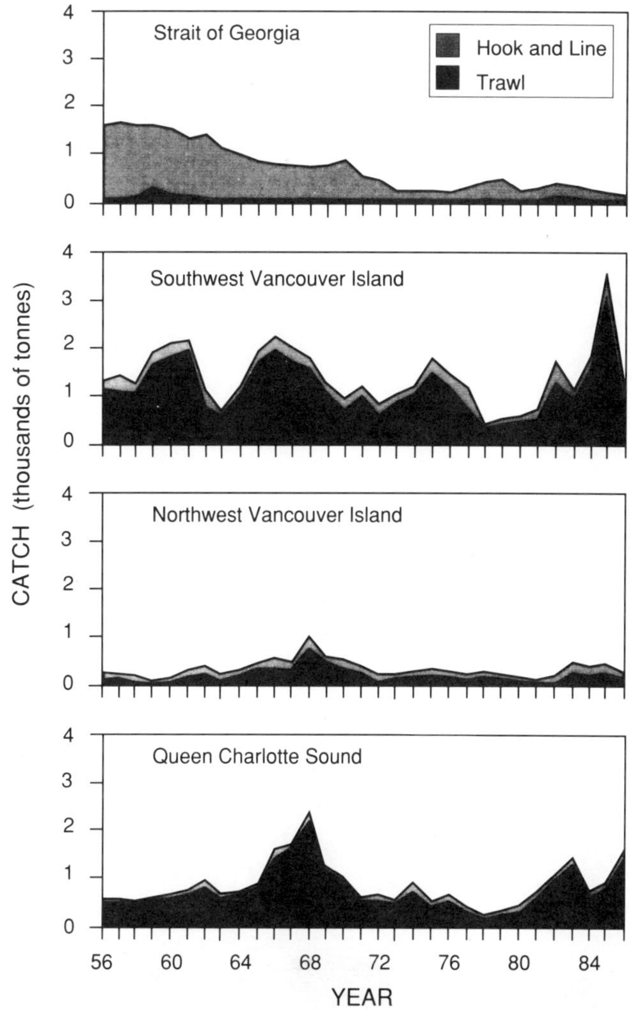

FIG. 7. Canadian catches of lingcod by area and fishing method off the west coast of Canada.

FIG. 8. Mixed species catch of groundfish, including lingcod, caught in the commercial trawl fishery off the west coast of Canada.

In the entire Strait of Georgia there are now less than 10 vessels that fish lingcod regularly.

Lingcod are also caught in less directed commercial fisheries. These have included longline, sunken gillnet, trap and seine fisheries. These fisheries have been solely Canadian except for incidental catches by the United

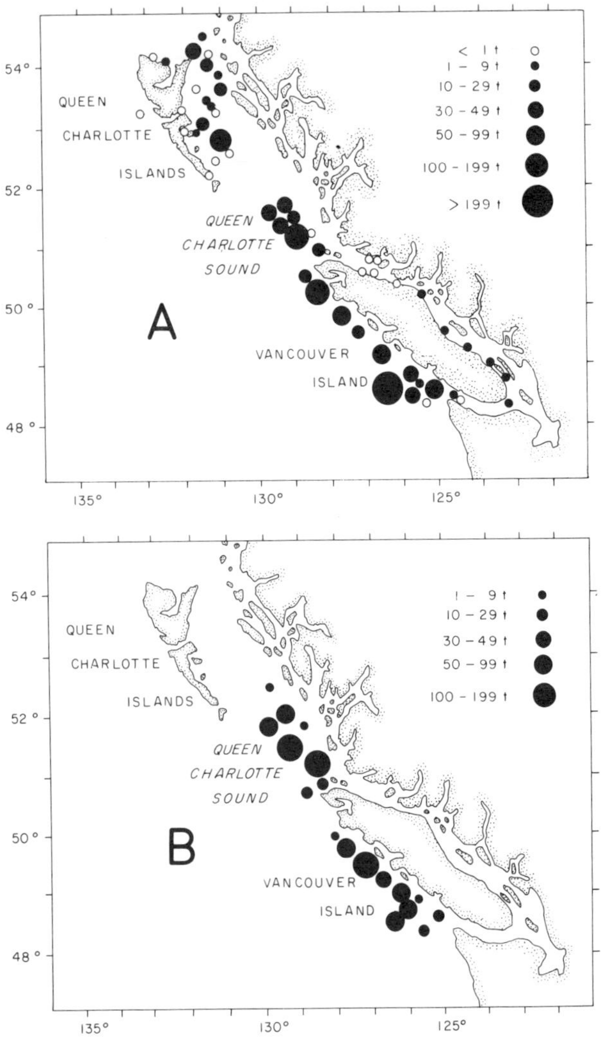

FIG. 9. Average geographic distribution of commercial catches in (A) the Canadian and (B) the U.S. trawl fisheries off the west coast of Canada during 1960–78.

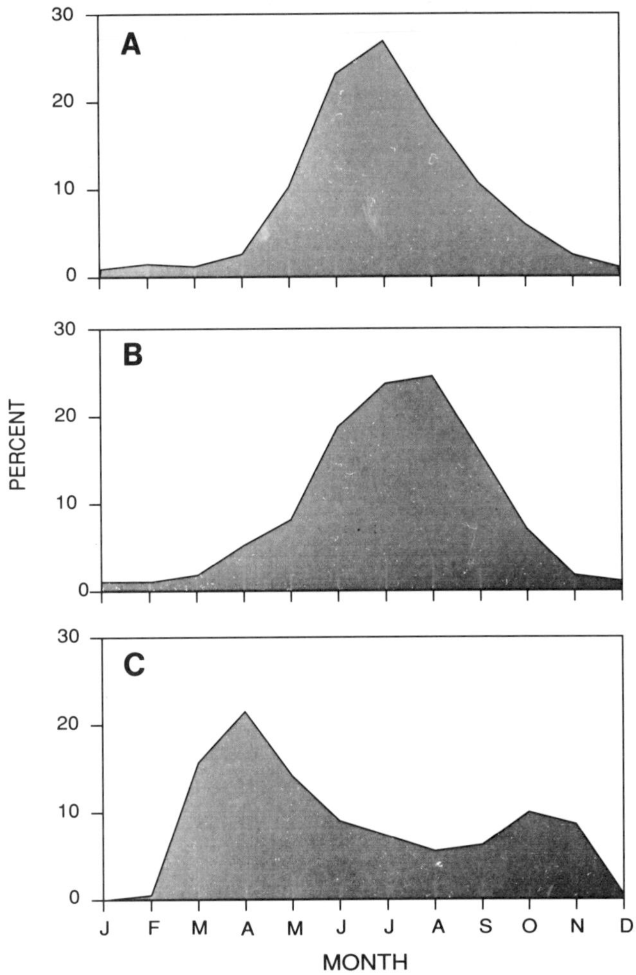

FIG. 10. Average monthly catches of lingcod from (A) the commercial trawl fishery off southwest Vancouver Island (1956–85) and (B) the sports (1980–85) and (C) commercial handline fisheries in the Strait of Georgia (1951–85).

States halibut longline and salmon troll fleets, and Japanese longliners fishing for sablefish. Catches of lingcod in these fisheries have been negligible compared to catches in the handline and trawl fisheries.

More recently, lingcod have been caught by recreational anglers and by scuba divers. It is widely recognized that sports fisheries underwent a rapid expansion during the 1960s, particularly in the Strait of Georgia. Unfortunately, lingcod catches by recreational anglers were not routinely estimated until 1980. The most reliable estimates are based on dock-side surveys to count the catch, coupled with aircraft surveys of the Strait of Georgia to estimate the number of boats fishing. Based on these surveys, recreational catches of lingcod accounted for an estimated 35% of the number of all lingcod landed and an average of 13% of the total all-species sports catches (including salmon), from the Strait of Georgia during 1980–85. During this period, an estimated average of 80 000 lingcod or 125 t/yr were landed by recreational fishermen. The annual trend in recreational catches of lingcod has varied widely with highs of 137 000 lingcod caught in 1980 and 1983. An independent survey of recreational fishermen based on less rigorous sampling methods indicated that about 90% of the recreational catch comes from the Strait of Georgia.

Recreational catches of lingcod from the Strait of Georgia were largest near Campbell River, accounting for 28% of the 1980–85 catch. Other areas of large catches were Pender Harbour (18%), the Gulf Islands (18%) and Victoria (15%). Unlike the commercial line fishery in the Strait of Georgia that catches most fish in the March–May period, catches in the recreational line fishery are largest during June–September (Fig. 10).

The increased popularity of scuba diving in the 1960s and 1970s resulted in a new underwater spear fishery. A recent survey of scuba divers in the Strait of Georgia indicated lingcod is the principal fish species sought by spearfishermen and that catches are significant. Conservative estimates of annual scuba catches were 80 t or 50 000 lingcod.

General Biology of Lingcod

Distribution and Movements

Lingcod are found only off the west coast of North America (Fig. 11). They are distributed in the nearshore waters from Baja, California, to the Shumagin Islands, with the centre of abundance off the coast of British Columbia (Fig. 12). They are commonly found on the ocean bottom at depths ranging from 3 to 400 m, however most occur in rocky areas from 10 to 100 m. Lingcod are abundant in the Strait of Georgia, off the west coast of Vancouver Island and in Queen Charlotte Sound.

Adult lingcod remain close to the areas where they spawn. Tagging studies show that after maturity they remain close to the reef or rocky area to which they recruited. Approximately 30 000 tagged lingcod have been released in Canadian and United States waters, of which 27 000 were from Canadian tagging programs (Appendix Table 2). As of December 1988, 5 000 had been recovered.

Tagging studies off the west coast of Vancouver Island show the vast majority (95% of recoveries in the first and second year after tagging, for example) stayed within 10 km of home, with only very slight movement beyond 50 km. Most of these offshore fish travelled in a southerly direction (Fig. 13), and it appears unlikely that offshore stocks mixed with stocks in the Strait of Georgia. Similar studies within the Strait support the conclusion that there is no mixing with offshore stocks. Scientists conclude that tagged lingcod tend, with occasional exceptions of travel up to 370 km, to remain within their release area. Lingcod are, therefore, considered to be a non-migratory species.

In our studies off the west coast of Vancouver Island, 7 429 fish were tagged in 1982 and 2 476 were recovered between 1982 and 1986 (Appendix Table 3).

In our studies in the Strait of Georgia, 10 293 lingcod were tagged between 1982 and 1985. The location of 722 recaptured fish was reported as of December 1986 (Appendix Table 4).

Juveniles disperse over a wider range than adults. Although only a few tagged juveniles were recovered, there was a general increase in the distance travelled after 2 years.

As was found for adults, there have been some unusually long migrations of tagged juveniles. For example, one juvenile released off the west coast of Vancouver Island was recovered off Oregon for a net movement of 510 km.

Reproduction

In Canadian waters, spawning begins in early December and continues through March (Fig. 14). Most of the spawning occurs from mid-January to mid-February (Appendix Table 5).

Preferred nest sites are located in crevices in rocky areas where there are strong currents. In the Strait of Georgia, the preferred nesting sites are located in current-swept narrows. Lingcod have been reported to select sites in shallow water where surface waves and tidal changes create circulating water movements. They rarely spawn in areas where tidal currents are weak.

The most important nesting habitat observed during our scuba surveys off the west coast of Vancouver Island was the exposed coastal reefs, located within 1–2 km of the coast. An estimated average density of 12 nests per 10 000 m^2 was observed in this area. In tidal swept passages or bays we estimated densities of nests to be fewer than 1 nest per 10 000 m^2. Observations by scuba divers

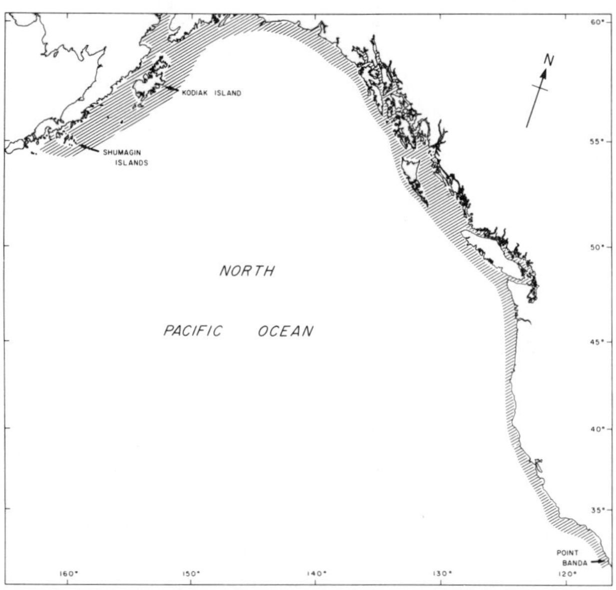

FIG. 11. Distribution of lingcod.

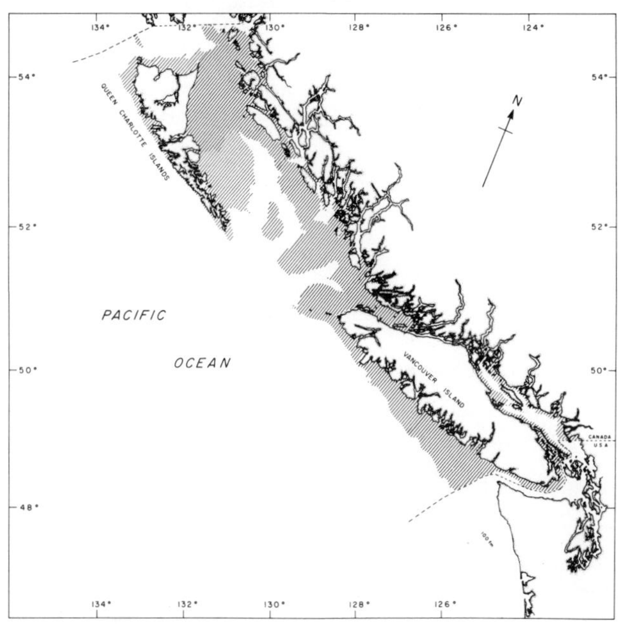

FIG. 12. Distribution of lingcod off the west coast of Canada.

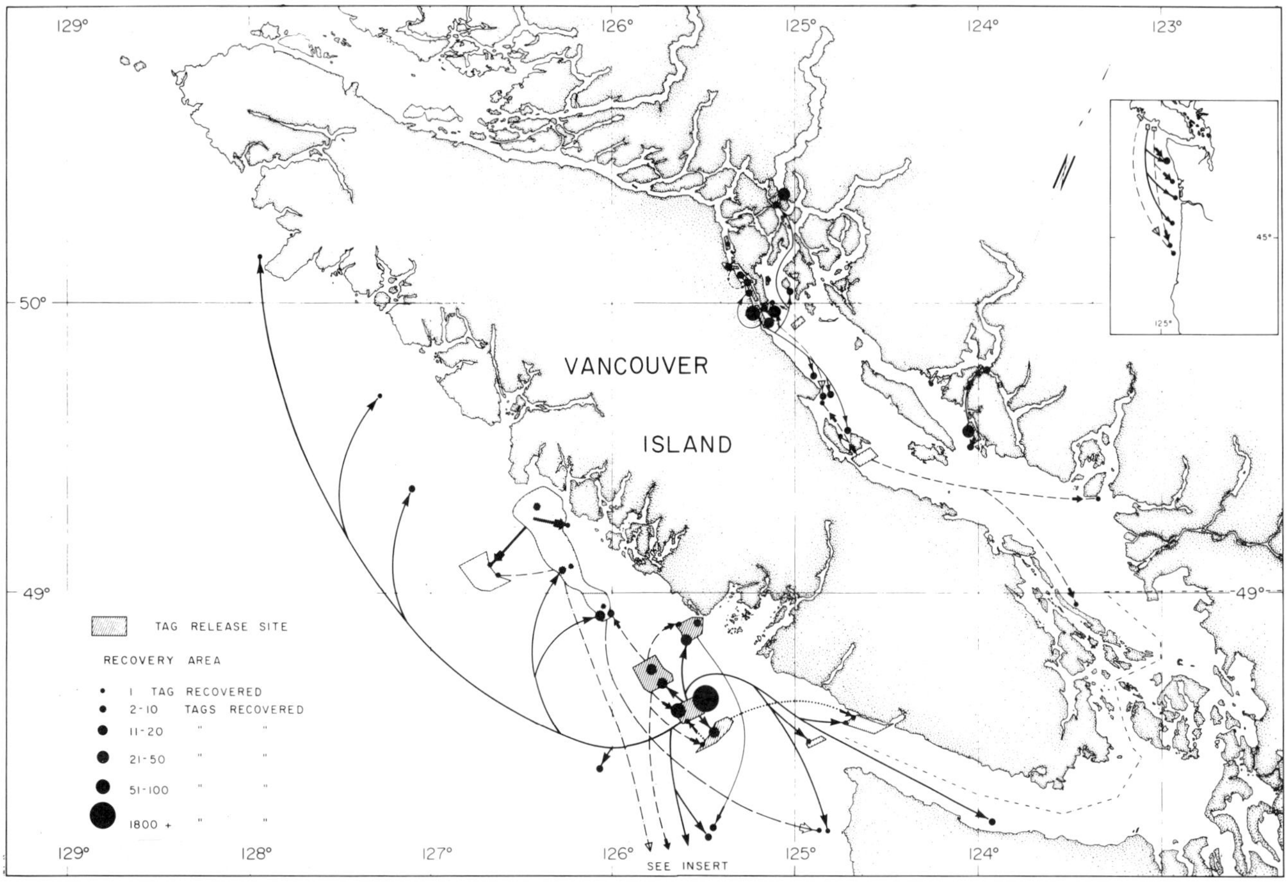

FIG. 13. Movements of tagged lingcod released off Canada during 1978–83 and recovered by December 31, 1983.

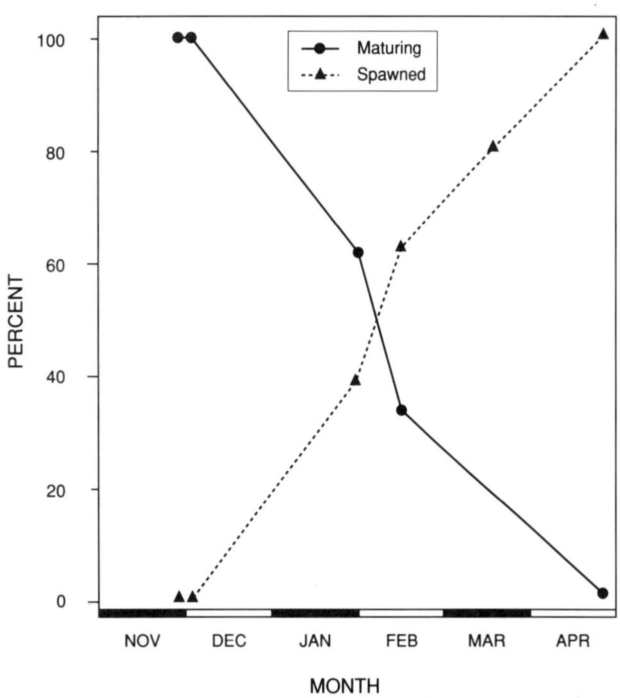

FIG. 14. Percent maturing and spawned female lingcod in trawl catches off the southwest coast of Vancouver Island during November 1976–April 1977.

were not possible in waters deeper than 30 m. However, in the Strait of Georgia, where sampling was conducted using hook and line gear, spawning males were found at 30–60 m.

Male and female lingcod appear on shallow nearshore trawling grounds from May to September and are almost entirely absent from these grounds during the winter, indicating a migration prior to spawning. The few catches in the winter were in deeper water (Fig. 15) and were almost entirely females. Because lingcod are not caught in bottom trawls in the winter, they are presumed to move to shallow rocky areas at this time.

The sex composition of commercial catches provides some indication when this segregation occurs. For example, the percentage of females in bottom trawl catches during 1977–80 in the summer averaged 57% while during the winter the percentage of females averaged 86%. The periods of segregation and desegregation occur approximately in October and April, respectively, suggesting that males begin to aggregate in October (Fig. 16) and desegregate after guarding the nest. In the Strait of Georgia, males find suitable nesting sites by December. At the time of the pre-spawning aggregation, males become noticeably territorial. During this period, when they are approached by scuba divers they will slowly circle the area around a nesting site rather than leave.

Males are in spawning condition earlier than females and move into the spawning area first, with larger males

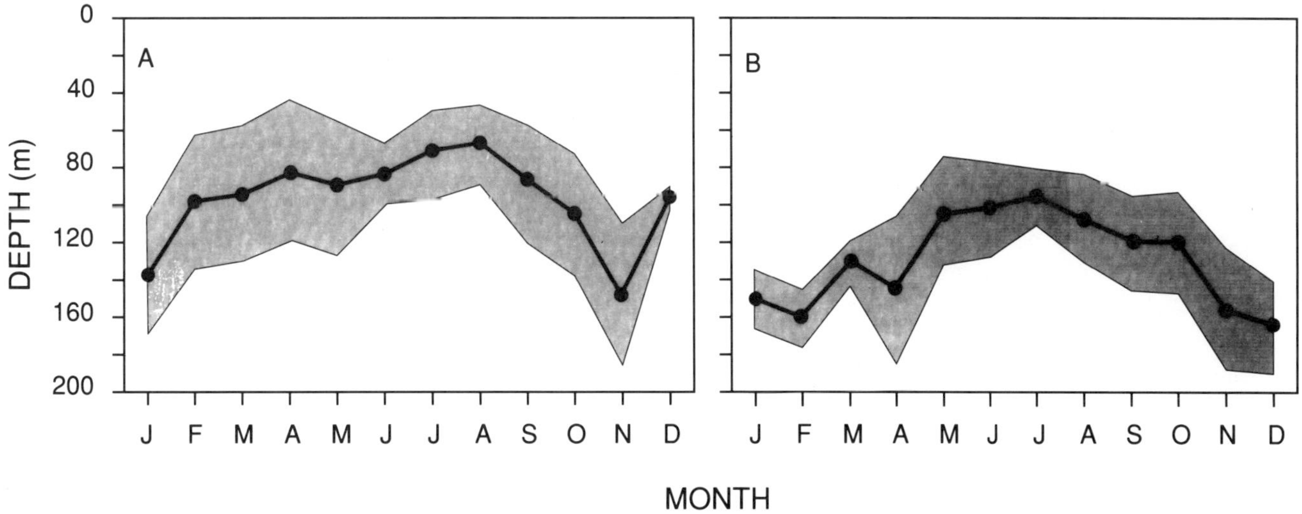

FIG. 15. Average monthly depth of trawl catches of lingcod (A) off southwest Vancouver Island and (B) in Queen Charlotte Sound, 1970–80. The shaded region represents ± one standard error.

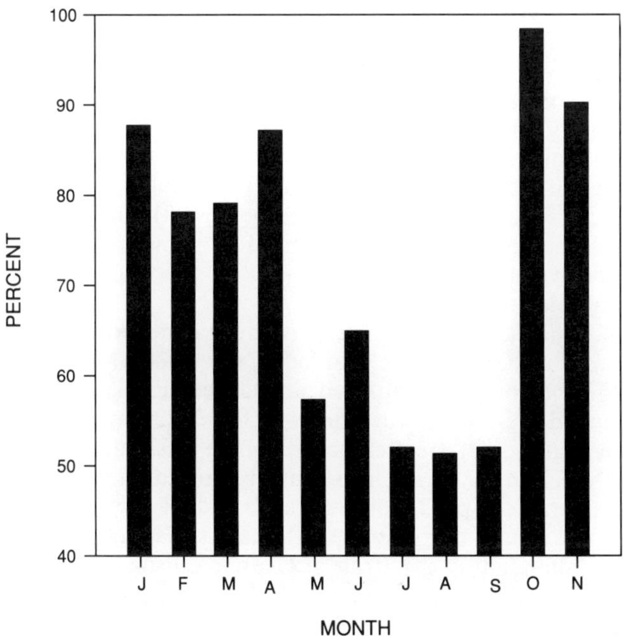

FIG. 16. Percentage of female lingcod in samples from the commercial bottom trawl fishery off the west coast of Canada by month during 1977–80.

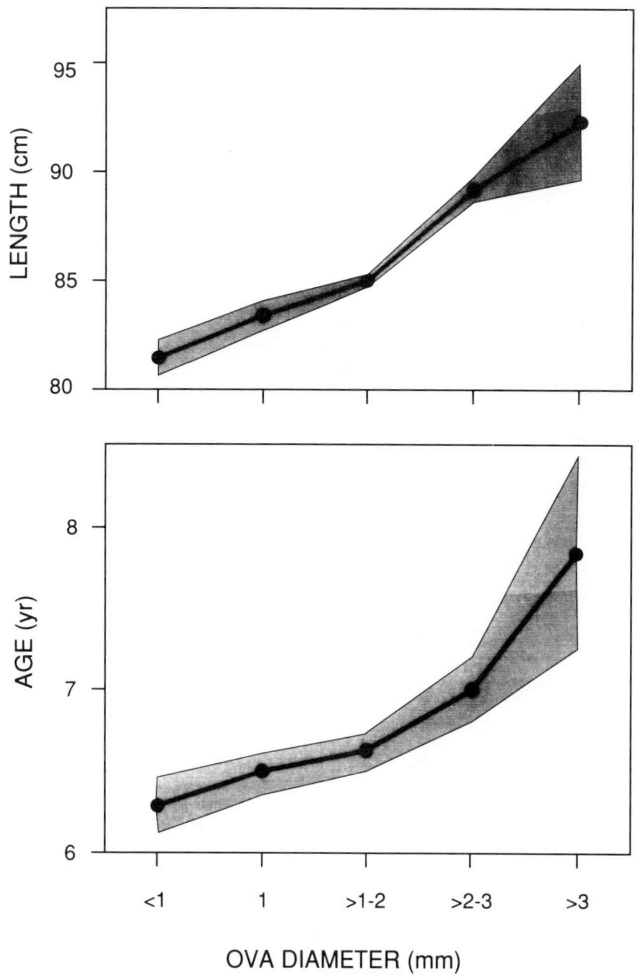

FIG. 17. Relationship between stage of maturity and mean length and mean age of maturing female lingcod caught off the west coast of Vancouver Island during research trawling in November-December 1977. The shaded region represents ± one standard error.

selecting the most suitable sites. It appears that larger and older females spawn first (Fig. 17).

Estimates of the mean age of mature female lingcod range from 3 to 5 years (Fig. 18) with a mean size of 61–75 centimetres (cm) (Fig. 19). Most males mature at age 2 and a length of 50 cm.

Lingcod probably spawn at night. The presence of a male at a nest site appears to be a stimulus to spawning. When spawning occurs, the female extrudes several layers of ova along with a gelatinous substance. The male then fertilizes the eggs. The female returns to add more eggs to the mass. This process continues until spawning is completed. Females leave the nest site after spawning. In most cases only one nest is deposited for each guarding male.

Studies at the University of Washington have found that after extrusion of the eggs along with a viscous gelatinous matrix, the eggs become firmly cemented to each other within 24–48 hours. Spaces exist between the hard-

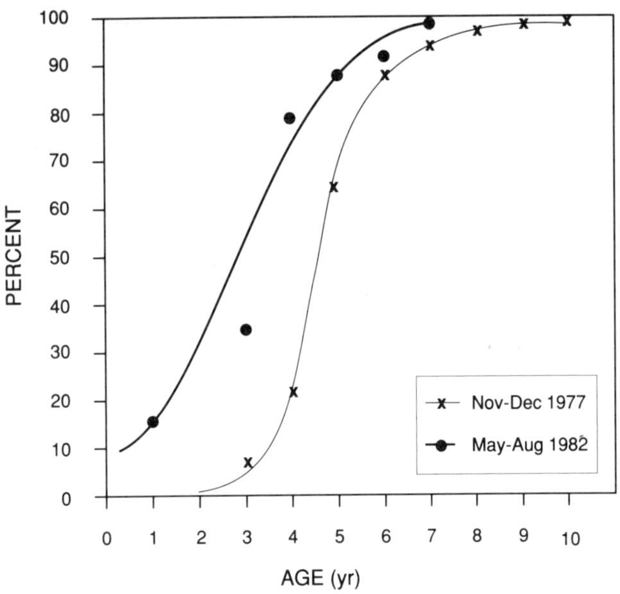

FIG. 18. Relationship between percent maturity and age of lingcod caught off the west coast of Vancouver Island during research trawling in November–December 1977 and May–August 1982.

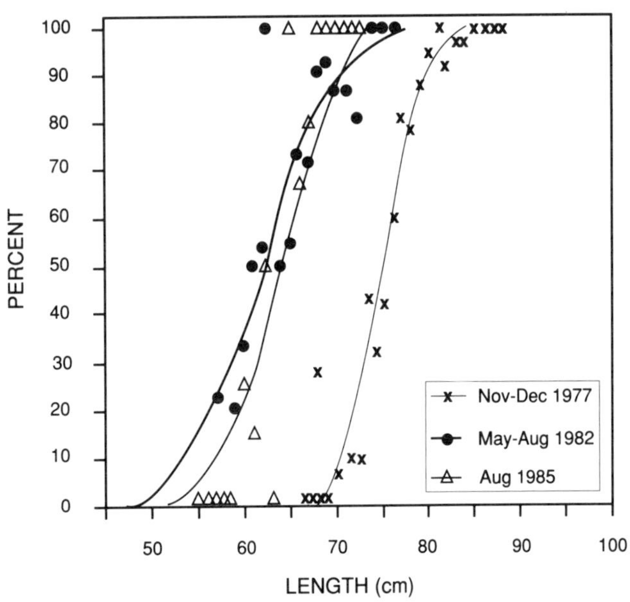

FIG. 19. Relationship between percent maturity and length of lingcod caught off the west coast of Vancouver Island during November–December 1977, May–August 1982 and August 1985.

ened eggs, presumably in the location where the gelatinous matrix existed. Eggs are about 2.8 mm in diameter when released and when fully hydrolyzed about 3.5 mm in diameter. Nest sizes of 3–65 litres have been reported. A fish 70 cm in length will produce about 97 000 eggs, and one 120 cm, about 490 000 eggs. The relative fecundity of lingcod (numbers of eggs per unit weight of fish) is approximately 26 eggs per gram of body weight. This is relatively low compared to other marine fishes but high among species that exhibit parental care.

The location of egg masses in areas receiving strong current ensures that the centre of the mass will have sufficient oxygen. Laboratory experiments showed that at a velocity of 10 cm per second (cm/s), oxygen concentration within the egg mass was nearly uniform and only 5–10% below the oxygen saturation of the ambient water. At velocities of 7.5 cm/s, oxygen was not distributed evenly throughout the egg mass. In velocities that are commonly found at lingcod nesting sites, the oxygen concentrations within the egg mass averaged 69% saturation. The mean oxygen saturation inside the egg masses that have been found in reduced current areas can be as low as 16% saturation. Weak currents result in egg mortalities in the interior of the egg mass or near the base of the egg mass. As eggs develop, the dead or necrotic region of the egg mass expands from the centre toward the periphery. In a Puget Sound study, egg masses in reduced current areas had egg mortalities ranging from 5 to 95% with a mean of 59%. With proper circulation, in the range of 10–15 cm/s, the rate of development was not retarded in the interior of the egg mass and no concentrations of dead eggs were found. Embryos in nests in high current areas were all in the same stage of development. It is apparent that the lingcod's method of spawning requires nest sites with adequate current velocities.

After spawning, male lingcod position themselves a distance of 1 m or less from the egg mass. Males have been observed to guard more than one egg mass, but this occurs only when nests are less than 1 m apart. It is unknown if these egg masses are from one or two females. Juvenile males have been found in nesting areas and it is possible that they also guard nests. Although aggressive behaviour has been reported by some, we did not observe any aggressive behaviour by guarding males in any of our studies.

In studies where males have been removed from the nests, some nests were guarded by new males. Most nests (67%), however, remained unguarded. Predators consumed these unguarded nests within 2–22 days, a clear indication that the presence of a male is essential for the protection of eggs. Unfortunately, guarding males are vulnerable to predation from seals and sea lions, and are easily caught by fishermen. Because seals and sea lions are abundant in the Strait of Georgia during the nesting period, we believe they are a major source of mortality of adult male lingcod at this time. However, it is important to remember that lingcod and sea lions co-existed long before we exploited lingcod.

In our Strait of Georgia study, 70% of the 77 nests observed were guarded. In another study, about 90% of the 39 most frequently observed nests in Puget Sound were guarded. However, off the west coast of Vancouver Island, less than a third of the 139 nests observed were found to have males nearby. We believe this indicates that males were frightened by the scuba divers rather than that the nests were unguarded. During this study, lingcod were particularly wary and it is probable that many unguarded nests would be guarded after the scuba diver left.

Scuba surveys in the Strait of Georgia indicate that larger nest guarding males were in deeper water. During hook and line surveys in the northern Strait of Georgia nesting males found in deeper areas were larger than those in shallow areas (Fig. 20). For example, the average length of males at depths ranging from 5 to 25 m was 57 cm. At depths ranging from 26 to 40 m, males averaged 66 cm in length. At depths greater than 40 m, males averaged 71 cm. Nesting studies in Puget Sound found that

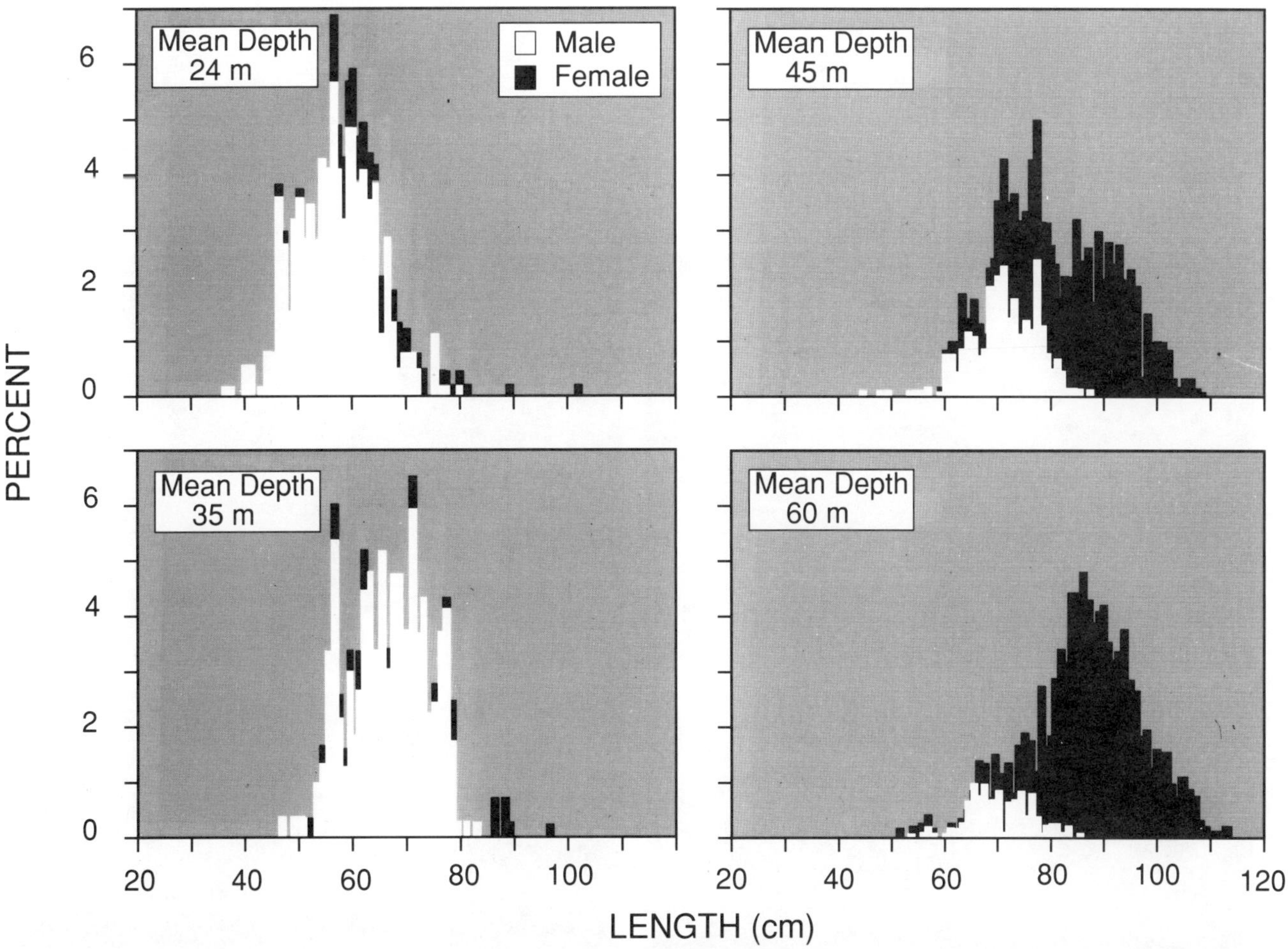

FIG. 20. Size distribution by sex and depth of lingcod caught during hook and line surveys in the Strait of Georgia during November 1982–March 1983.

shallow water nests were more often unguarded than deep water nests. This suggests that there may be some differences in guarding effectiveness related to size and that egg survival is potentially lower in shallow water nests. Apart from mortality due to predation, there is some indication that reduced egg viability occurs in nests in shallow water. Anecdotal information suggests that nests in shallow water in the Strait of Georgia were more abundant in past years than in recent years. Although highly speculative, this would suggest that in periods of high abundance, for example the 1920s, shallow water spawning habitat may be utilized to a greater extent.

Nests are eaten by both fish and invertebrate predators (Fig. 21 and Appendix Table 6). The main predators in the Strait of Georgia study were kelp greenlings (*Hexagrammos decagrammus*) and striped seaperch (*Embiotoca lateralis*). Greenlings and seaperch both preferentially attacked unguarded nests and were instantly chased away from guarded nests, while invertebrate predators and small fish such as sculpins were ignored.

The incubation period ranges from 5–11 weeks and averages about 7 weeks in both Puget Sound and the Strait of Georgia. Hatching success varies considerably, ranging from 10–27% in our studies in the Strait of Georgia to 90% in Puget Sound. The majority of nests that hatch do so by mid-March or April. However, some nests lasted as late as June and there is some indication that poor ventilation is responsible for protracted incubation.

Hatching begins from the outside of the egg mass and progresses inwards. Hatching in high current areas occurs quickly with most larvae emerging in a few hours. In the Puget Sound study, hatching in the laboratory, under conditions of high current velocity, was usually completed within 24–48 h. Most larvae emerged in a few hours and the first larvae to hatch were significantly larger than those that hatched from eggs near the centre. In our studies of egg masses in the wild, most of the eggs hatched in high current areas within 24 hours. Other studies have reported hatching times of 3–7 days (Appendix Table 5).

The guarding male remains close to the nest during hatching. Smaller fish such as sculpins and striped seaperch have been observed feeding on newly hatched larvae. The mortality due to immediate predation of newly hatched lingcod is unknown. However, the presence of guarding males appears instrumental in reducing larval mortality. Nest guarding males usually leave the nesting site soon after hatching. In fact, our studies indicate males abandon nests before all eggs hatch and that egg viability is reduced in nests that hatch late in the season.

As mentioned, lingcod nests require a suitable flow of water on the exterior to ensure that there is sufficient

FIG. 21. Potential predators of lingcod nests: (A) yelloweye rockfish, (B) quillback rockfish, (C) copper rockfish, (D) cabezon, and (E) sunflower star feeding on lingcod eggs. (Photos courtesy of Rick Harbo, Department of Fisheries and Oceans, Fisheries Branch, Nanaimo, B.C.)

the necessary water flow, nests need to be deposited in areas where currents are strong. These areas also contain many potential predators. In summary, the biology of lingcod is uniquely adapted to spawning in these highly productive areas by producing adhesive eggs, having high fecundity (compared to other species that exhibit parental care), and large egg size, a nest guarding behaviour, and synchronous hatching of all eggs in a nest.

In general, fish that release eggs directly into ocean currents produce more eggs per unit weight of fish than fish that provide some form of protection for eggs. This method of spawning, while requiring large numbers of eggs, insures some eggs hatch in a favourable environment and that eggs are dispersed throughout the range of the species. Species such as Pacific halibut (*Hippoglossus stenolepis*), sablefish (*Anoplopoma fimbria*) or Pacific hake (*Merluccius productus*) disperse eggs by depositing them directly into the ocean and these species have broad ranges of distribution. Deposition of eggs in one large nest means that dispersion only occurs after hatching. This method of reproduction provides recruitment primarily into the immediate vicinity of the spawning area. In the next section the limited dispersion of juvenile lingcod is described. This phenomenon indicates that colonization and recruitment occur in localized areas. This, in turn, means that local areas can be overfished and that local stocks need to be carefully managed.

Early Life History

Larval lingcod hatch at a total length of 6–10 mm (Fig. 22). They are relatively well developed with distinctive spots of dark pigment dorso-anteriorly and bands of dark pigment on each side of the dorsal fin and lateral line (Fig. 23). Except for these areas, the body is relatively unpigmented. Larval lingcod at this stage are easily distinguished from other fishes by the large yellow oil globule in the gut region and the relatively large gape of the mouth.

The transition from larvae to juveniles is a gradual process. Initially, the larvae are relatively passive components of the plankton community. As they grow, they become increasingly proficient at swimming. By early April, young lingcod are about 20 mm in total length (Fig. 24). At this time, they are bright green dorsally and silver ventrally. By late April, lingcod in surface waters are 30 mm in length and have begun to assume the adult appearance. At this time the larvae are concentrated in

FIG. 22. Lingcod hatching after 5-11 weeks of embryonic development. (Courtesy of Rick Harbo, Department of Fisheries and Oceans, Fisheries Branch, Nanaimo, B.C.)

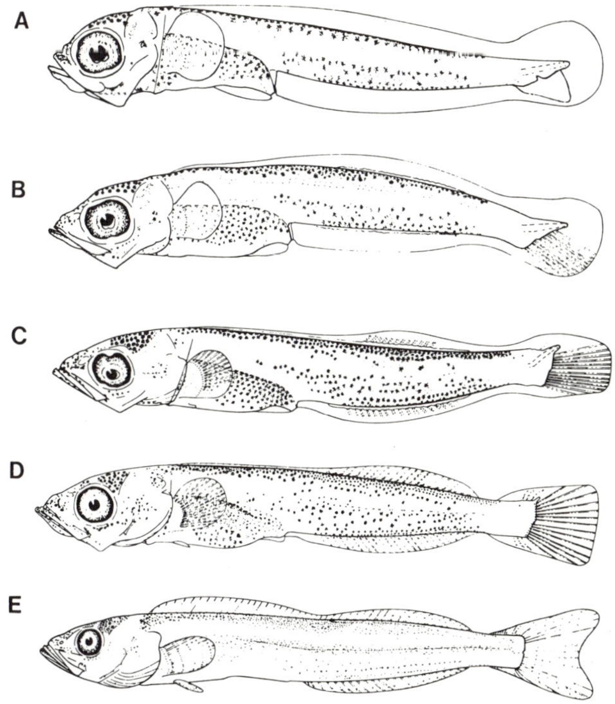

FIG. 23. Larval lingcod at different life stages. (A) 9.8 mm, (B) 12.2 mm, (C) 15.4 mm, (D) 19.0 mm, and (E) 36.0 mm. (Source: Kendall and Vinter 1984).

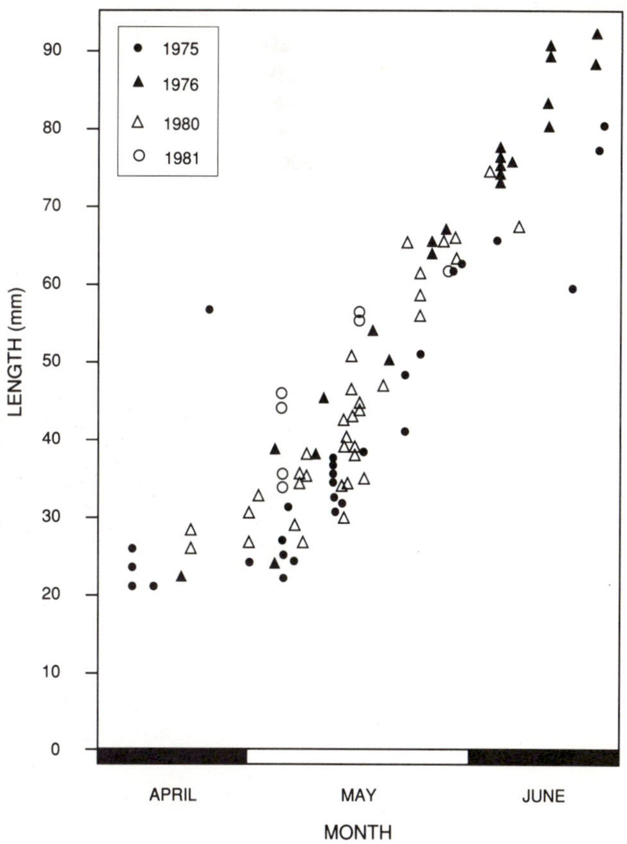

FIG. 24. Growth of larval lingcod in research purse seine catches from the Strait of Georgia during April–June 1975-76 and 1980–81.

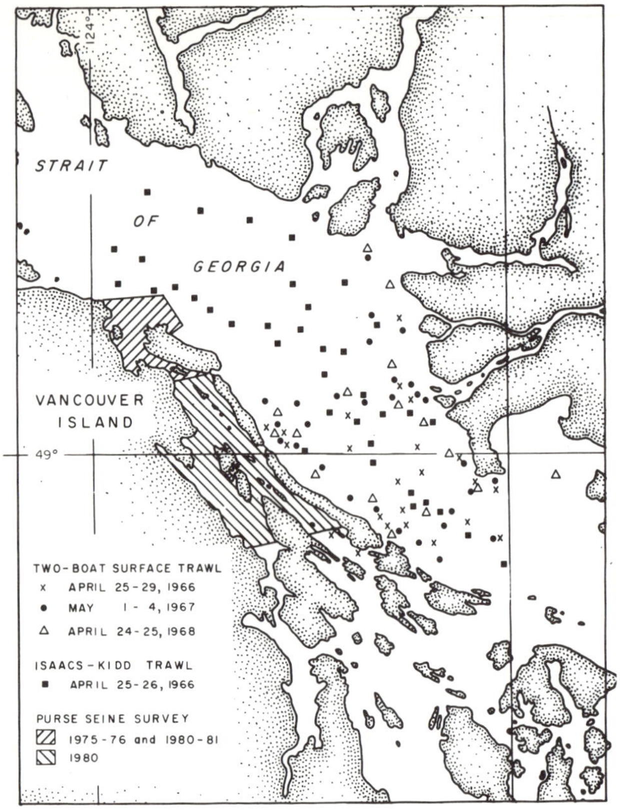

FIG. 25. Research survey sites for larval lingcod in the Strait of Georgia.

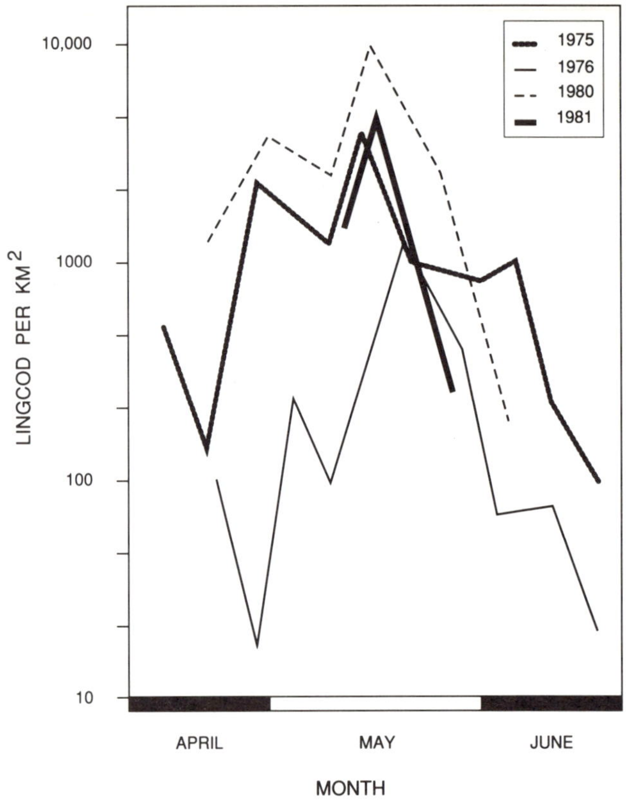

FIG. 26. Densities of larval lingcod estimated from purse seine surveys in the Strait of Georgia during April–June 1975–76 and 1980–81.

the upper 3 m of the water column during the day; they disperse or migrate to deeper depths at night.

Lingcod have been captured throughout the open waters of the Strait of Georgia (Fig. 25), however they were more abundant along the eastern shore of the Gulf Islands, where adult populations are abundant. The relative abundance of larval lingcod varied considerably between surveys conducted in 1966–68, 1975–76, and 1980–81. The range in densities was 900 to 10 000 larvae per square kilometre. Largest densities during the mid-May period of maximum concentration occurred in 1980. In 1975, 1980, and 1981, the relatively large abundance of larvae estimated in surveys in the Strait of Georgia resulted in a large number of age-1 lingcod.

The timing of the accumulation of larvae in the surface waters and their subsequent decline was similar in 1975–76 and 1980–81 (Fig. 26). During 1975–76 and 1980–81, larval densities increased during late April and mid-May, and declined dramatically by early June. From late May to early June, larvae settle to the bottom as juveniles when they are about 70–80 mm (Fig. 27). The disappearance of larval lingcod from the surface waters coincides with an increase in the presence of juvenile lingcod on bottom near kelp or eelgrass beds.

FIG. 27. Juvenile lingcod soon after taking up bottom life. (Courtesy of Rick Harbo, Department of Fisheries and Oceans, Fisheries Branch, Nanaimo, B.C.)

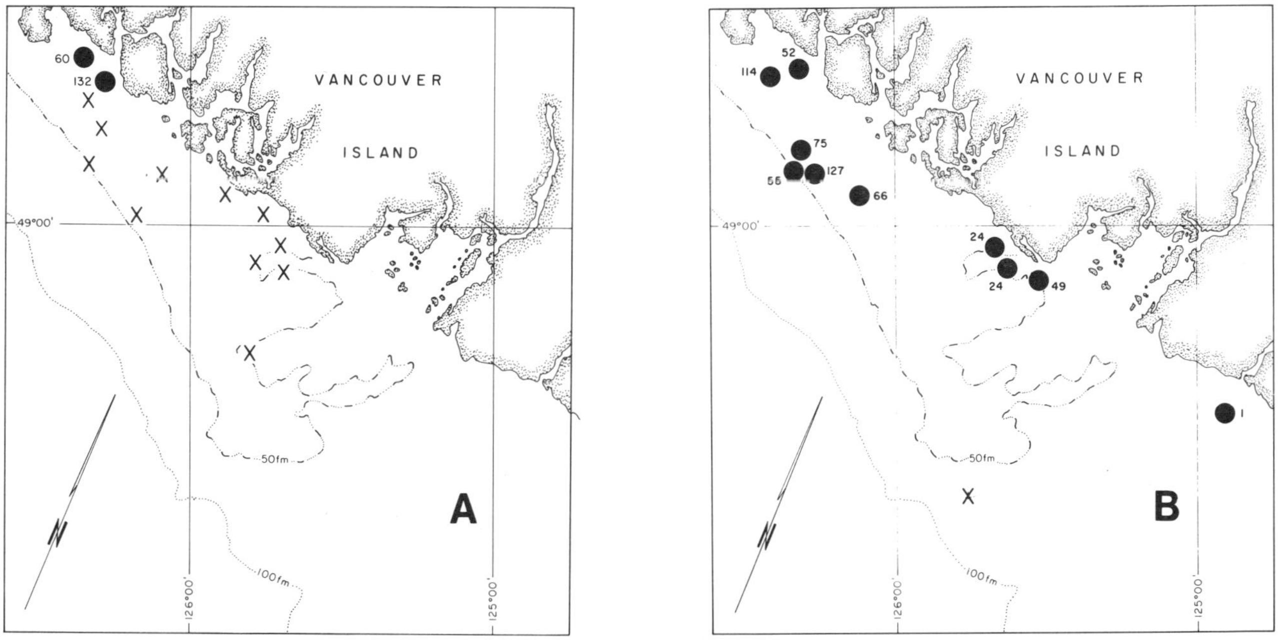

FIG. 28. Number of age-0 lingcod caught per hour in (A) July and (B) September 1978 during research trawling off the southwest coast of Vancouver Island. Sites with zero catches are shown as an "X".

FIG. 29. Young age 2 or 3 lingcod in typical hard bottom habitat. (Courtesy of Rick Harbo, Department of Fisheries and Oceans, Fisheries Branch, Nanaimo, B.C.)

Surveys conducted off the west coast of Vancouver Island indicate that juvenile lingcod move from inshore areas to a wider range of flat bottom areas by September (Fig. 28). In the Strait of Georgia young-of-the-year lingcod are also found on flat bottom in August, in areas that are not typical habitat of older lingcod. By age 2, lingcod begin to move into habitats of similar relief and substrate as older lingcod, but they remain at shallower depths (Fig. 29).

Food and Feeding

Larval lingcod have a large mouth and are capable of eating larger organisms than other fish of the same size. For example, a 40 mm lingcod can ingest food items that cannot be consumed by salmon until they are at least 60 mm. During the initial stages of larval development, the diet consists of small to medium sized copepods. As growth continues, there is a gradual transition in size and

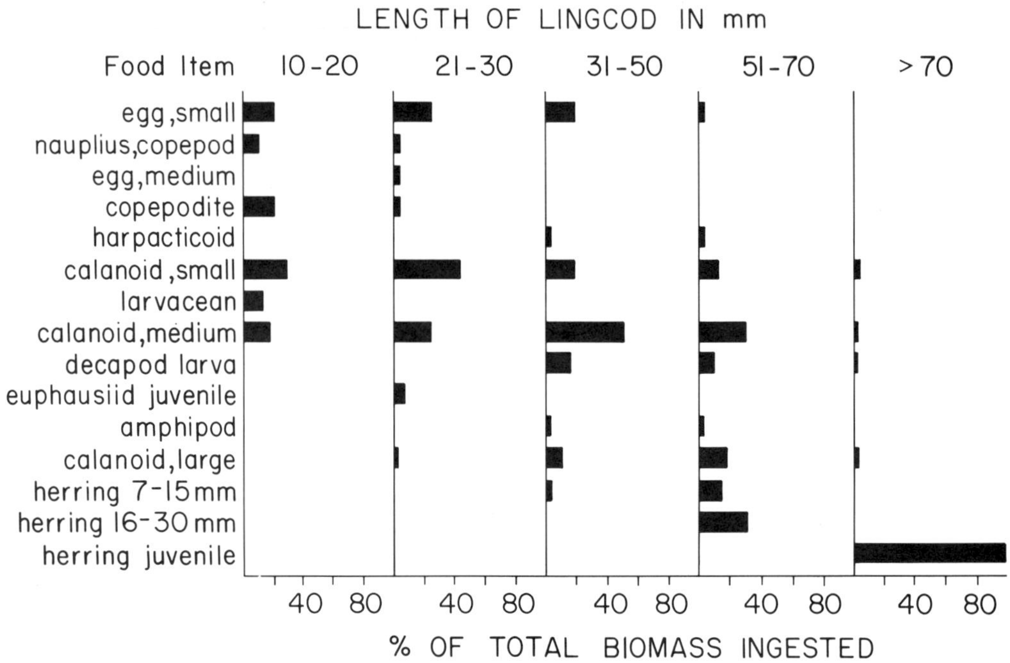

FIG. 30. Size and diversity of food items ingested in relation to larval lingcod size.

diversity of food items (Fig. 30). Larger food items include large copepods, crab larvae, amphipods, euphausiids and larval herring. Just prior to the transition to bottom life, at a length of about 70 mm, lingcod diet shifts exclusively to juvenile Pacific herring (*Clupea harengus*). This transition occurs over a period of about 1 week.

Herring continues to be an important food item for juveniles. Other food of juveniles consists of Pacific sand lance (*Ammodytes hexapterus*), flatfish (Pleuronectidae), shiner perch (*Cymatogaster aggregata*) and juvenile walleye pollock (*Theragra chalcogramma*). Invertebrates such as shrimp (*Neomysis macrops*) and prawns (*Pandalus danae*) are consumed but are less important. Although we have not found lingcod in the stomachs of juvenile lingcod in the wild, cannibalism has been reported for juvenile lingcod held in captivity.

Adults are voracious predators of fish and invertebrates. Many species have been found in the stomachs of adult lingcod including Pacific herring, Pacific sand lance, flatfish, rockfish (*Sebastes*), spiny dogfish (*Squalus acanthias*), young lingcod, Pacific cod (*Gadus macrocephalus*), Pacific hake (*Merluccius productus*), sablefish (*Anoplopoma fimbria*), Pacific tomcod (*Microgadus proximus*) and salmon (*Oncorhynchus*). Species of fish most often found in the stomachs of adults are herring and hake. Invertebrates observed in the stomachs of adults are crabs, shrimps, squid and octopus.

Commercial hook and line fishermen often use live bait, such as Pacific herring, greenlings (Hexagrammidae), shiner and pile perch (*Rhacochilus vacca*) and lingcod skins to catch lingcod. This use of lingcod skins supports our laboratory observations that lingcod readily feed on other lingcod.

Predation

Little is known about predation on lingcod. During a study of salmon abundance near the Qualicum River in 1981, 9 out of 121 coho salmon (*Oncorhynchus kisutch*) contained larval lingcod. As mentioned earlier, it is known that adult lingcod are eaten by marine mammals. They have been observed in the stomach contents of sea lions and harbour seals from British Columbia waters. Fishermen and scuba divers frequently report seeing sea lions feeding on lingcod, suggesting that marine mammals are an important source of predation particularly during spawning. Once lingcod mature, their large size, territorial habit and voracious appetite make them less vulnerable to predation by other fish. Therefore, except for predation by marine mammals, predation is unlikely to be an important source of mortality among older lingcod.

Age and Growth

In 1977, a method of ageing lingcod was developed that used thin sections of fin-rays. The ability to age lingcod was a major breakthrough in our understanding of lingcod biology, allowing estimates of age-specific mortality, growth rates and the number of eggs (fecundity) produced for each age group. Fins can be easily recovered from commercially caught fish without affecting market value. This method was subsequently shown to be accurate for all age groups using fish that were tagged and injected with oxytetracycline, a substance that marks bones with a fluorescent dye (Fig. 31). During periods of slow growth in the winter, the bone is more compact and appears as a narrow light zone in the dried tissue when viewed in a microscope with transmitted light. The fluorescent dye showed that these light zones formed once a year. Thus, the age of a fish can be determined simply by counting these lines. The fin-ray method of age determination can also be used for juveniles but the lines are less distinct in young fish. Therefore, it is better to use length because the rapid growth during this period

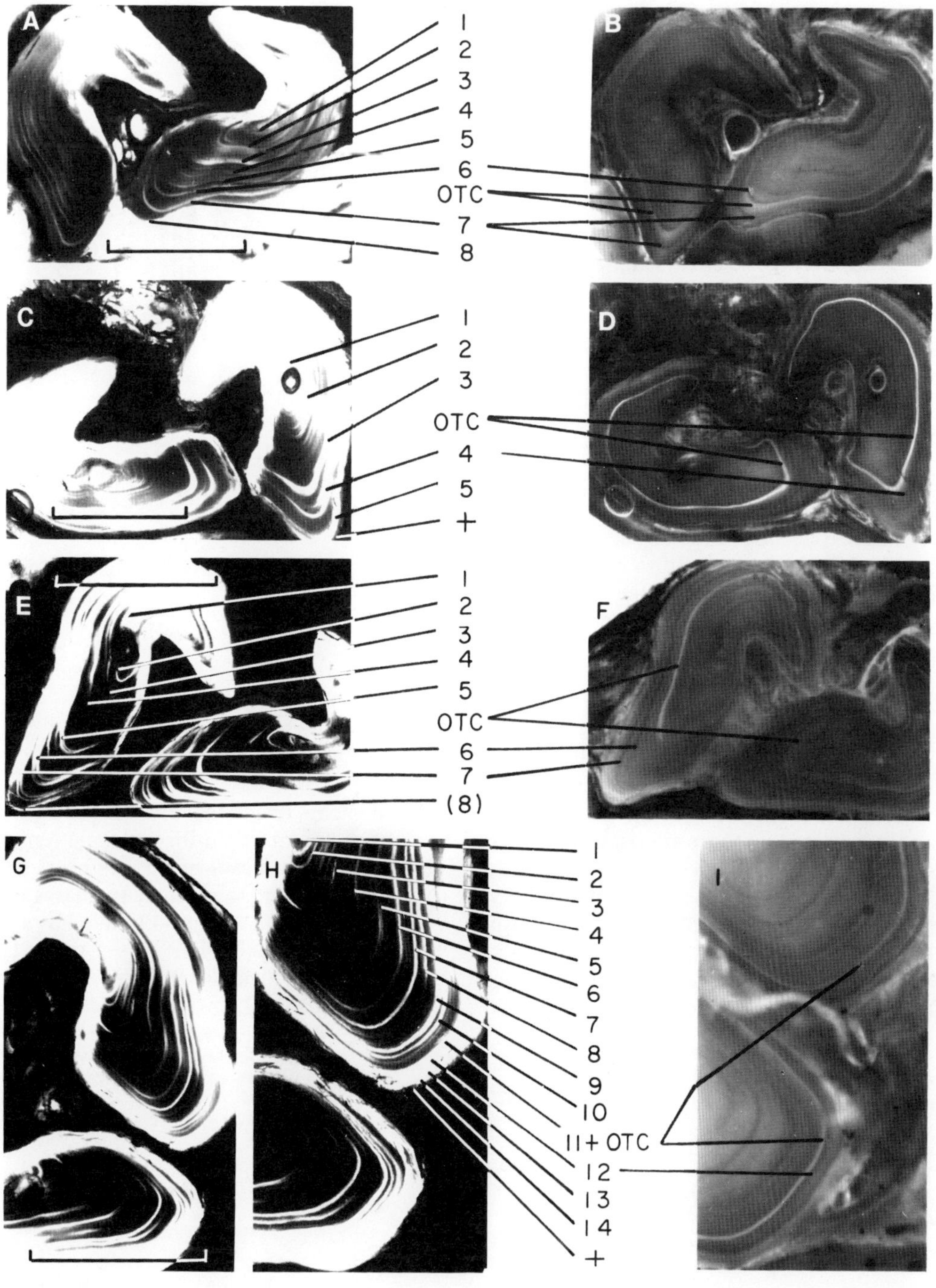

FIG. 31. Cross sections of fin-rays from lingcod that were tagged, injected with oxytetracycline and recovered after more than 1 year of liberty. The oxytetracyline mark was photographed using reflected UV light (Fig. 31B, D, F, I) and the annulus pattern of opaque and translucent zones (Fig. 31A, C, E, G, H) was photographed using transmitted tungsten light. The bar represents 1 mm. OTC = oxytetracycline. Age in parentheses indicates that a new annulus appears to be forming on the edge, but the fish was recaptured prior to January 1. A–B are from a 69 cm length male; C–D are from a 58 cm male; E–F are from a 60 cm female; and G is from a 77 cm male. H and I are enlargements of G.

produces distinct length groups that do not overlap with younger or older fish. After age 2 there is an overlap in lengths of age groups and length becomes unreliable as an indicator of age.

Lingcod increase in length rapidly during the first few years of life. During the larval stage daily growth is 1.6 mm. By the time they are 8 months old (age 0+) they average 21 cm in length. By the time they are 20 months old (age 1+) they are an average 33 cm. By the following year (age 2+) they average 45 cm in length and weigh about 0.8 kg (1.7 lb). After age 2, females grow faster than males. For example, an age-3 female is an average of 3 cm larger than a male of the same age and an age-13 female is an average of 19 cm larger than a male of the same age. This difference in size increases after age 8, when growth rates of males become greatly reduced. Females, however, continue to grow rapidly until age 12–14 yr (Fig. 32).

The maximum age in our studies was 14 yr for males and 20 yr for females. The maximum length recorded in our studies was 123 cm and the largest fish ever reported was 150 cm in length and 32 kg (71 lb) in weight.

Lingcod growth is not the same in all areas (Fig. 33). In the Strait of Georgia and off the west coast of Vancouver Island growth is similar. Off the west coast of Vancouver Island males aged 3–8 yr increased an average of 3.6 cm/yr in length. Females of the same age increased an average of 5.2 cm/yr. Lingcod in Queen Charlotte Sound are larger than in southern areas. Differences in growth corroborate the tagging information reported earlier, indicating that there are separate stocks off Canada.

Growth in weight was not routinely measured directly in our studies because of the difficulty of weighing fish when at sea. However, weights can be calculated so that length can be converted to weight (Fig. 34). For example, an age-4 male off the west coast of Vancouver Island

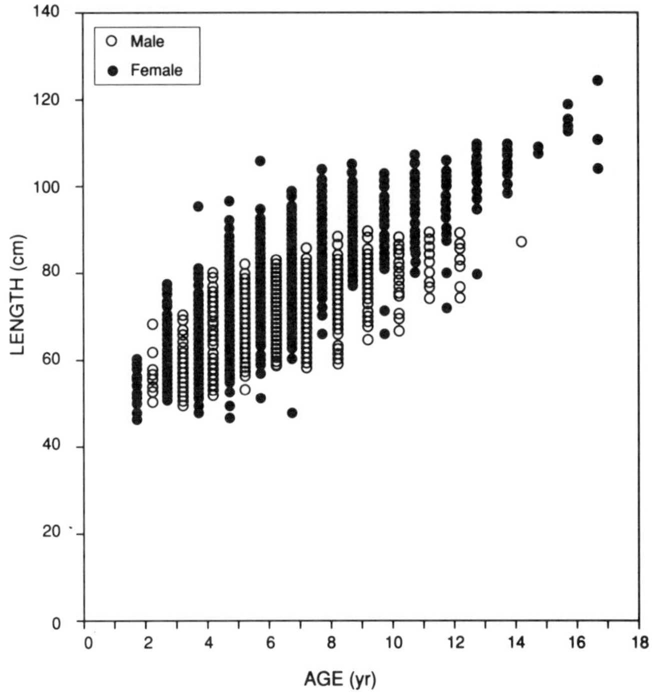

FIG. 32. Growth in length of male and female lingcod off southwest Vancouver Island showing the variation in length at age of individual fish.

with a mean length of 63 cm has an estimated mean weight of 2.4 kg or 5.3 lb. An age-4 female with a mean length of 68 cm has a mean weight of 3.1 kg or 6.9 lb. As fish get older, the weight increases exponentially in relation to length. An age-13 male has a mean length of 84 cm and a mean weight of 6.1 kg or 13.6 lb. An age 13 female has a mean length of 103 cm and a mean weight of 11.9 kg or 26.3 lb (Appendix Table 7).

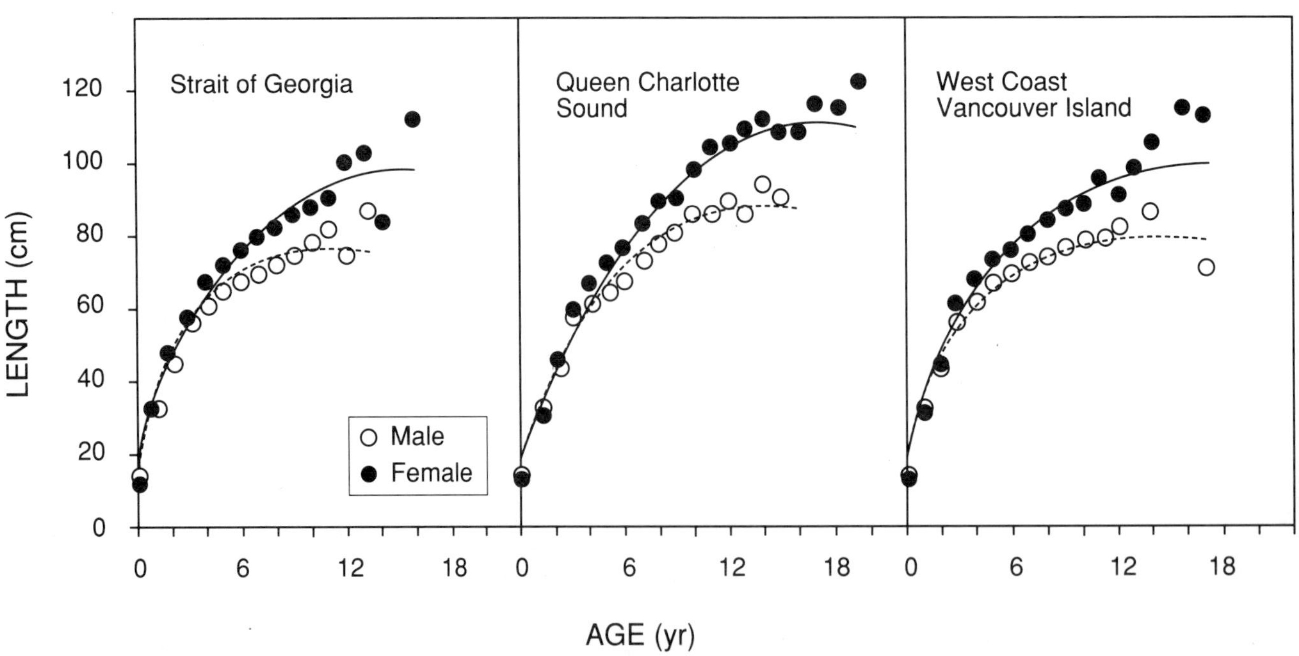

FIG. 33. Growth in length of male and female lingcod showing the variation in mean length at age among different geographic regions off Canada's west coast.

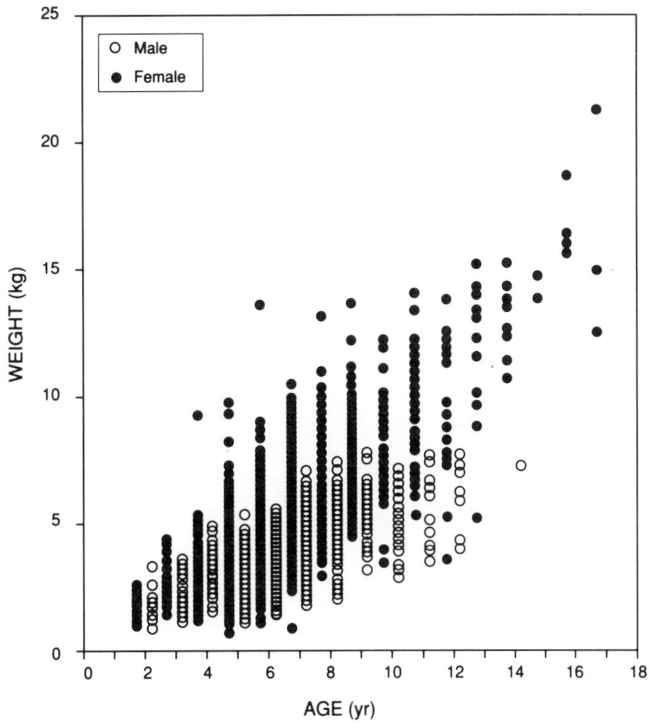

FIG. 34. Growth in weight of male and female lingcod off southwest Vancouver Island showing the variation in the estimated weight at age of individual fish.

Condition of Lingcod Stocks and Fisheries

Age and Size in Sports and Commercial Fisheries

Biological investigations of lingcod received considerable attention throughout the 1920s and 1930s. Despite the long history of study, the information necessary for assessing the condition of the resource has been slow to acquire. With the development of a proper age determination method in 1977, we now routinely assess the age composition of lingcod caught by fishermen. Estimates of the age composition are used to estimate survival rates, and examine the relationship between fishing rates and population growth, long term yield and reproductive potential.

Beginning in 1977, biological samples were collected from the major bottom trawl fisheries off southwest Vancouver Island and Queen Charlotte Sound. Samples from the Strait of Georgia commercial handline fishery were first collected in 1979. Biological samples from the recreational fishery in the Strait of Georgia, however, were not collected until 1983. Most age and size data are collected during the period of intensive fishing. For the trawl and the recreational fisheries, this occurs during the summer. Most commercial handline fishing in the Strait of Georgia occurs either in the spring or the fall.

The recreational and commercial fisheries catch different sizes of lingcod (Fig. 35). In the recreational fishery during 1983, 50% were less than 60 cm in length while nearly 100% of the lingcod in commercial samples were greater than 60 cm. Based on the relationship between length and age for younger lingcod, the recreational fishery catches age 2 to 4 lingcod. On the other hand, the larger size of lingcod in the commercial fishery is partially the result of the size limit of 58 cm in length for commercial fisheries. This size limit gives total protection to age 2 lingcod and partial protection to age 3 fish but no protection to fish older than age 4. Because female lingcod mature between the ages of 3–5 yr, the size limit of 58 cm provides only partial protection to immature females. However, because males mature at age-2 and a mean length of 45 cm, immature males are not retained in the commercial fishery. There are currently no size restrictions on the recreational fishery. If it can be demonstrated that survival of juveniles caught and released in the sports fishery is sufficiently high, it probably would be advisable to extend the size limit to all fisheries.

Lingcod are not fully recruited to the commercial fishery until age 6 (Fig. 36). Based on the relative abundance in samples collected during 1977–85, age 5 and age

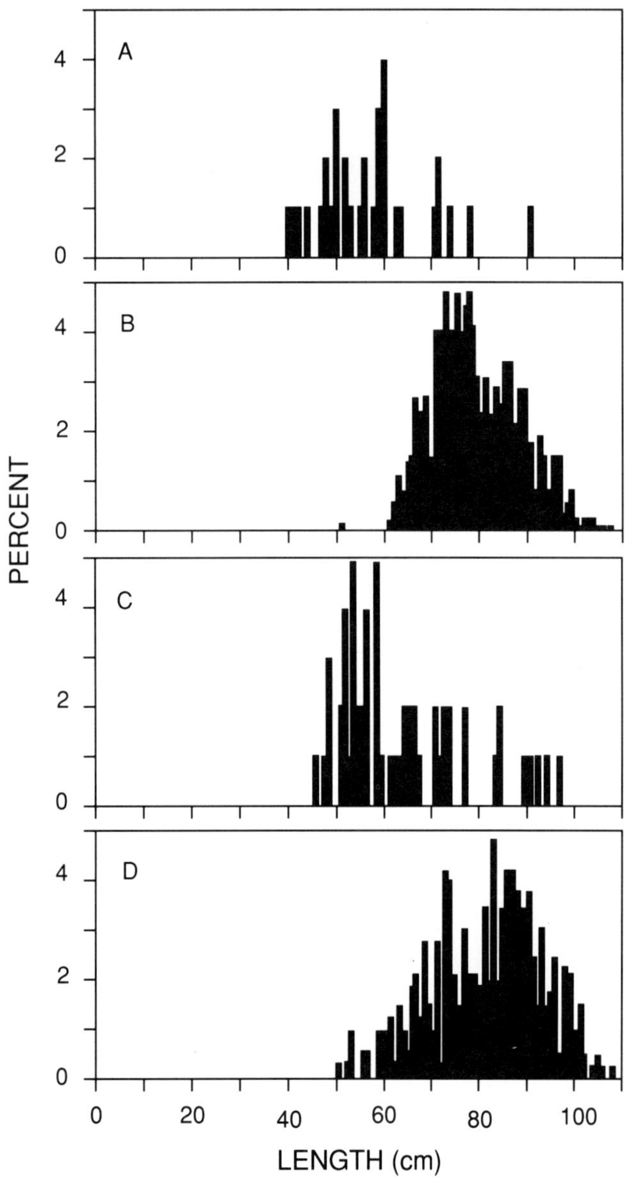

FIG. 35. Length distribution of lingcod sampled from A) the sports fishery and B) the commercial fishery compared to the length distribution from tagged lingcod recovered from C) the sports fishery and D) the commercial fishery in the Strait of Georgia in 1983.

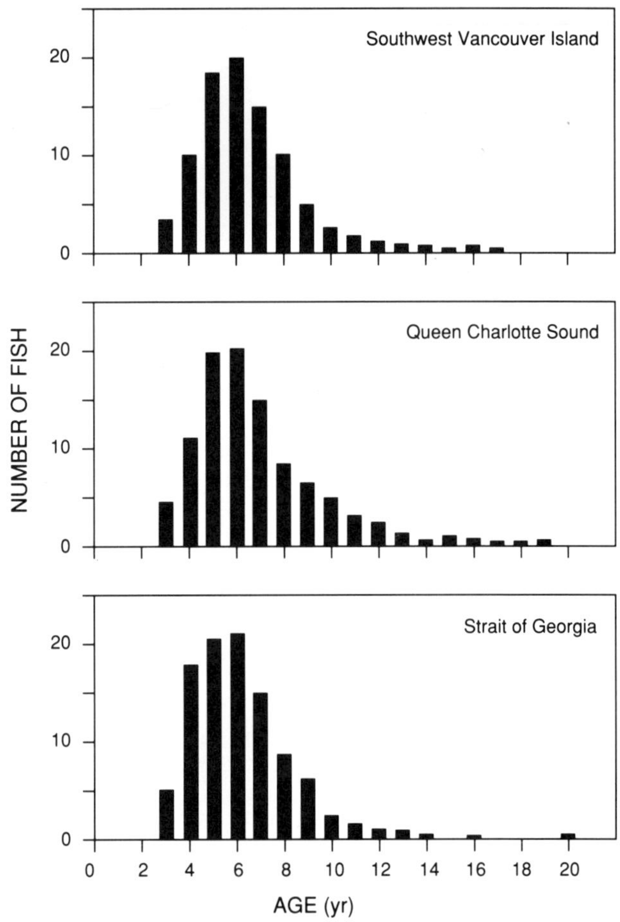

FIG. 36. Average age composition of lingcod in the commercial trawl fishery off southwest Vancouver Island (1977–83) and in Queen Charlotte Sound (1977–83), and the commercial handline fishery in the Strait of Georgia (1979–83).

6 lingcod accounted for an average of 19% and 20% of the landed catch, respectively. The relative contribution of older age-classes declines rapidly after age 6.

Every 2–3 years, large numbers of larval lingcod survive, producing large numbers of adults or what biologists call a strong year-class. In recent years, strong year-classes occurred in the offshore fishery in 1977–78 and 1980–82. The 1977–78 year-classes first appeared in the fishery in 1981 and have dominated the fishery since. In fact, in 1985, the 1978 year-class along with strong 1980–82 year-classes, contributed to record catches in the commercial trawl fishery off the west coast of Vancouver Island. The same strong year-classes were present in samples from the west coast of Vancouver Island (Fig. 37–38) and from Queen Charlotte Sound (Fig. 39–40), indicating that strong year-classes occur as a coast-wide event in all offshore regions off British Columbia.

Lingcod in the Strait of Georgia are also characterized by strong year-classes. Data collected from the handline fishery in the Strait of Georgia indicates the 1975–77 year-classes were dominant features in the fishery in 1981–82 (Fig. 41–42). In 1983, the 1979 year-class as age-4 lingcod dominated the catch. However, the strong 1977–78 year-classes seen in fisheries off the west coast of Vancouver Island and in Queen Charlotte Sound did not occur in the Strait of Georgia, confirming that factors that produce strong year-classes operate independently in the offshore and inshore areas.

It is not known if strong year-classes occurred before stocks were fished commercially, or if the fishing in some way produced cycles of abundance. The synchronous occurrence of strong year-classes throughout offshore areas, however, suggests that ocean conditions are a major cause in the production of strong year-classes. It is commonly believed that the availability of food immediately after hatching controls the abundance of a year-class. The productivity of the ocean and the currents can affect the availability of food for the recently hatched lingcod. However, predation or cannibalism could be an important source of juvenile mortality. Whatever the cause, strong year-classes are a feature of the population that affects the size of the catch in both the commercial and recreational fisheries.

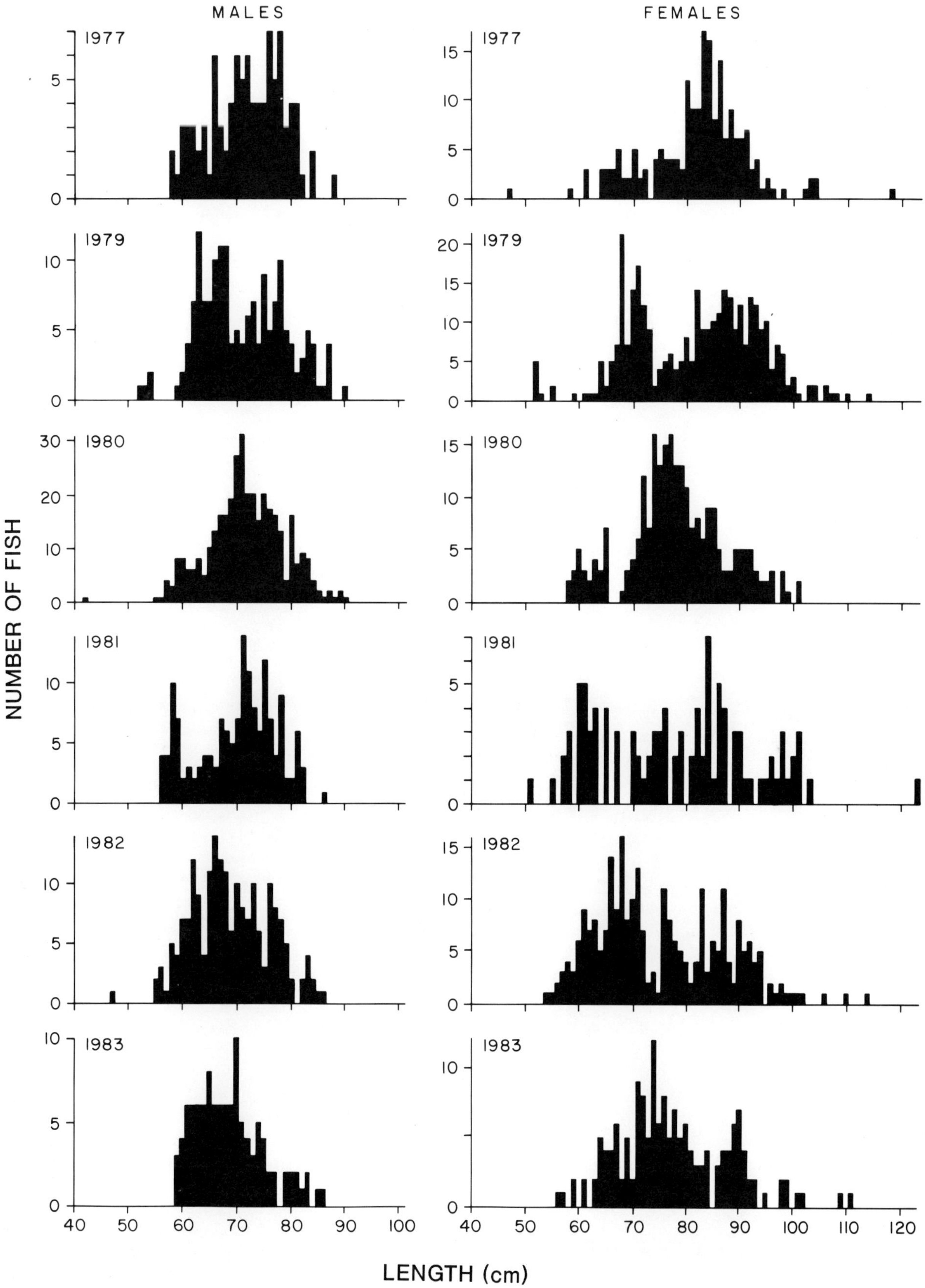

FIG. 37. Length distribution of male and female lingcod in the trawl fishery off southwest Vancouver Island during 1977–83.

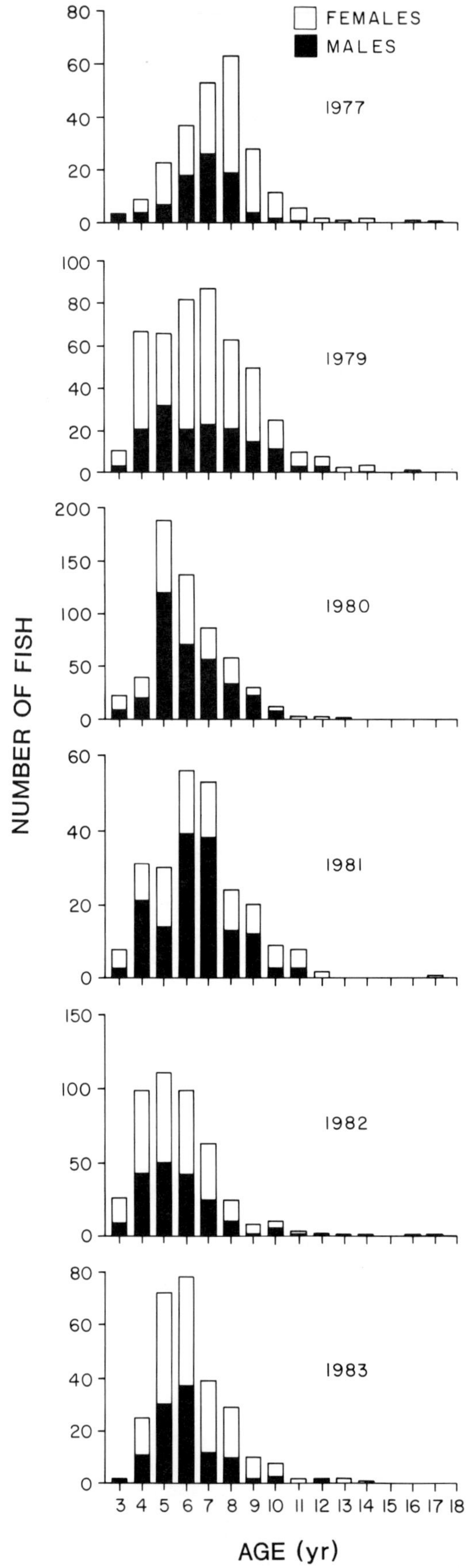

FIG. 38. Age distribution of male and female lingcod in the trawl fishery off Southwest Vancouver Island during 1977–83.

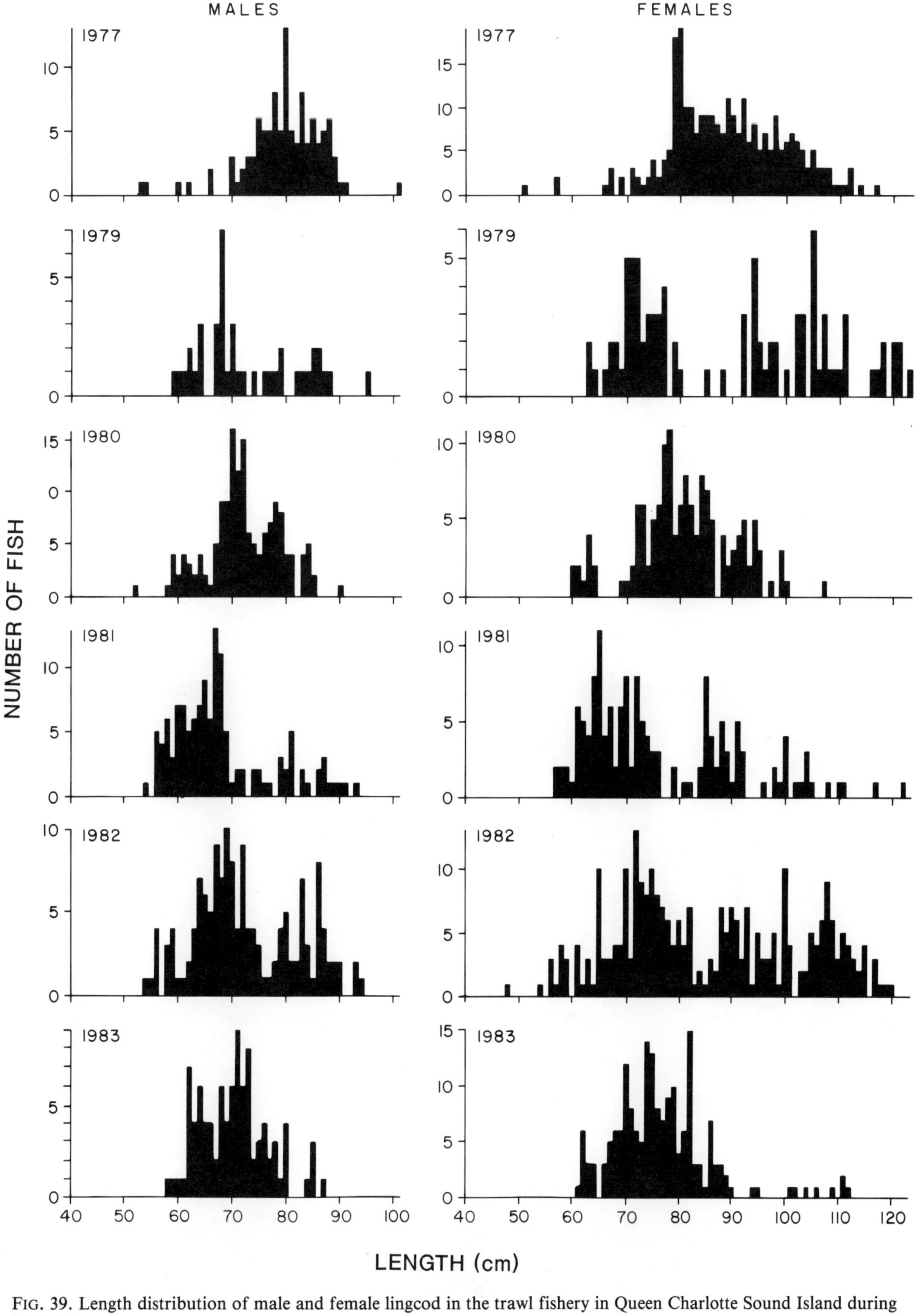

FIG. 39. Length distribution of male and female lingcod in the trawl fishery in Queen Charlotte Sound Island during 1977–83.

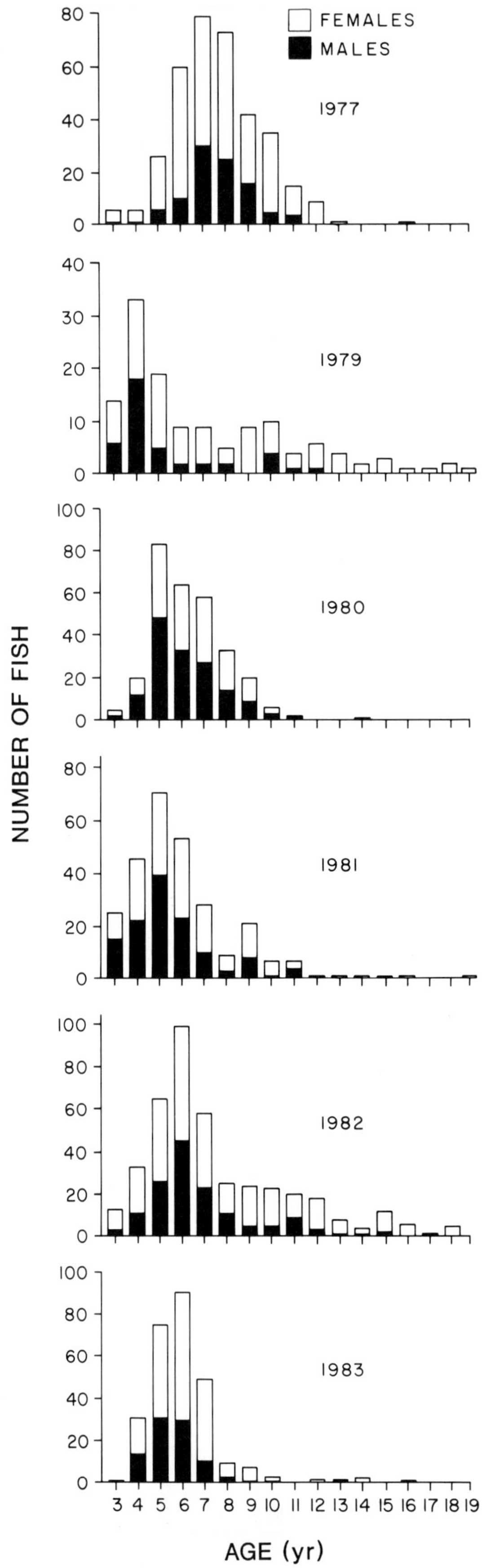

FIG. 40. Age distribution of male and female lingcod in the trawl fishery in Queen Charlotte Sound during 1977–83.

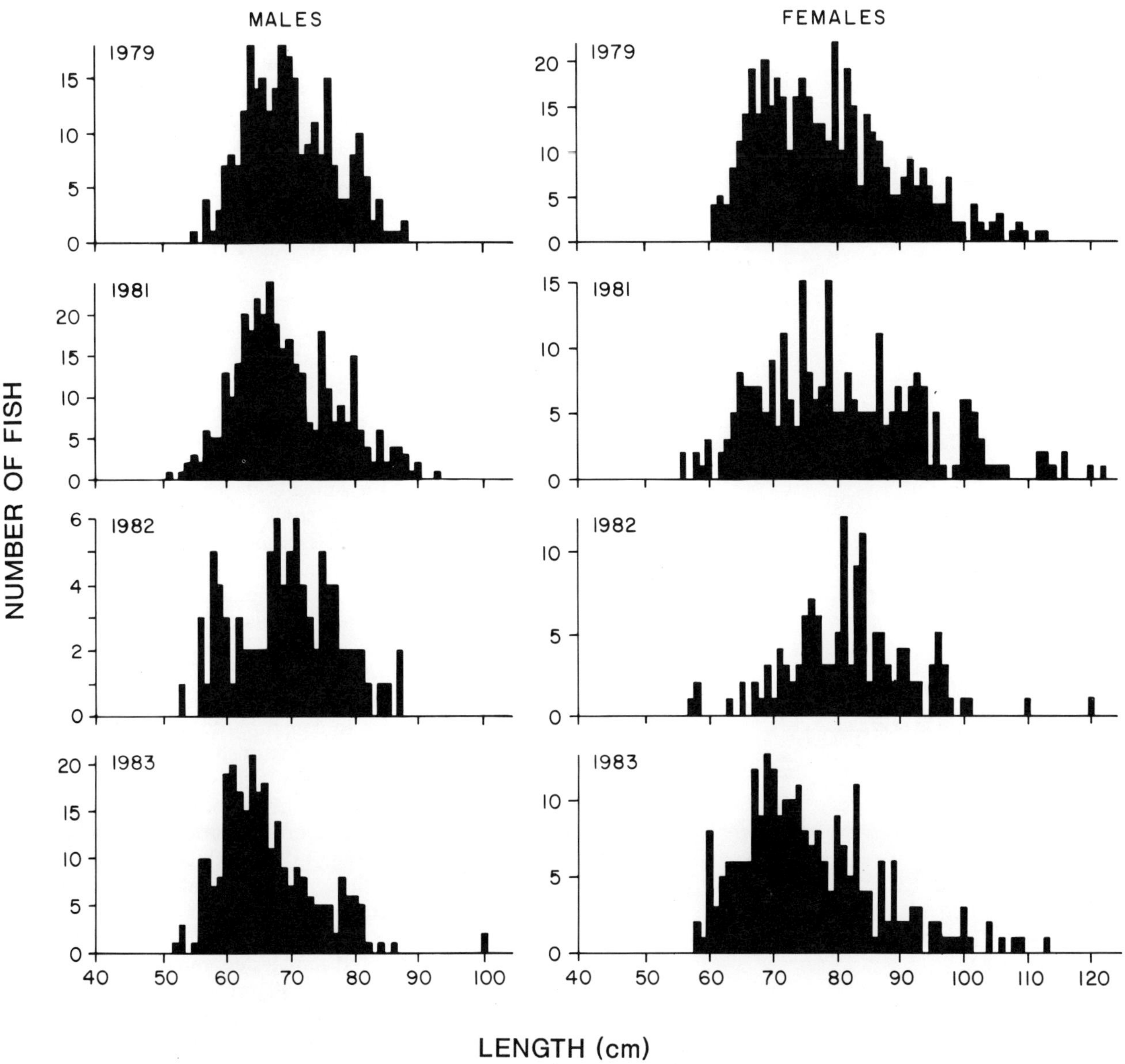

FIG. 41. Length distribution of male and female lingcod in the commercial handline fishery in the Strait of Georgia during 1979–83.

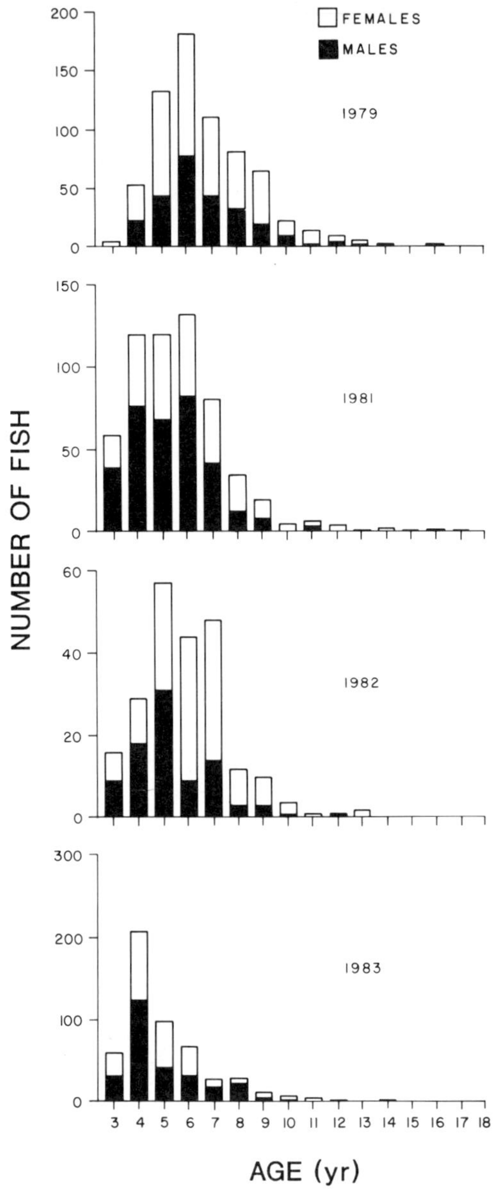

FIG. 42. Age distribution of male and female lingcod in the commercial handline fishery in the Strait of Georgia during 1979–83.

Survival and Abundance

The survival and abundance of fishes are influenced by the number of eggs produced and the cumulative effect of mortality throughout life. Survival is influenced directly by oceanographic factors, by predators, by food, and fishing. The impact of fisheries on survival is of particular interest to fisheries biologists. Determining the number of fish that can be harvested from a population requires an understanding of survival from natural causes and from fishing, the age at which the fish matures, the number of eggs produced and how the ocean environment affects the survival and growth of young fish. Also, it is important to remember that a fish has evolved to optimize its survival in the natural environment. When a species is fished commercially, the fish in the population begin to adapt to this new source of mortality. It is important to note that the changes that occur among commercially exploited species are only beginning to be understood by biologists.

In general, a species that grows quickly, has a comparatively low natural survival, and matures early is considered to be highly productive. Such fish can be harvested at a higher rate than less productive species. Lingcod are fast growing with a moderately low natural survival rate and an early age of maturity, thus are considered to be a productive species.

Estimates of survival can be calculated in different ways depending on the quality of information available. A necessary source of information is a time series of data that enables a comparison to be made of the abundance of a species at different time intervals. An example of such a time series is the relative abundance of one year-class or a cohort measured each year as it passes through a stock. In this case, a stock is defined as a geographically distinct group of fish, and a population is a group of interbreeding fish, usually composed of a number of stocks. Another index of abundance is the catch-per-unit-effort or CPUE. This is calculated by dividing the catch, say in a commercial fishery, by the amount of effort or fishing effort used to catch the fish.

A method commonly used by fisheries biologists to estimate survival is to compare the catch of a cohort at successive ages. Such a relationship is commonly called a catch curve. We used the catch of lingcod at age 6 to age 12 to estimate survival for stocks off the west coast of Vancouver Island, Queen Charlotte Sound and the Strait of Georgia (Fig. 43). The fluctuations in recruitment make it difficult, however, to identify the relationship between catch and abundance. For the age range of 6–12 years, the estimates of survival ranged from 0.52 to 0.68 for males and females combined. Survival rates for males (0.50–0.60) were slightly lower than for females (0.55–0.68), however, the difference is not statistically significant at the 95% confidence interval, so it is not certain that the differences in survival rates are real.

In addition, the partitioning of overall survival into components of survival from natural mortality and fishing mortality is important. Natural mortality results from predation, disease or even old age, while fishing mortality is caused by humans. Estimates of these two components of survival require more information than is obtained from the catch and age data. To partition the two kinds of mortality we use catch and effort data combined with information on growth rates and recruitment, and results from tagging experiments.

A simple tagging model can separate the two components of survival. The CPUE of tagged lingcod recaptured between 1982 and 1986 from releases off southwest Vancouver Island in 1982 was used to index abundance. The relationship between the CPUE and the catch in successive years of recapture was used to estimate survival from natural mortality and fishing. The model equation $I_t = \sigma (I_{t-1} - qC_{t-1})$ states that the CPUE I_t in year t, is equal to the CPUE I_{t-1} in the previous year minus the catch C_{t-1} of tagged lingcod times the natural survival σ. The units of catch are scaled to CPUE rather than abundance by the catchability coefficient q.

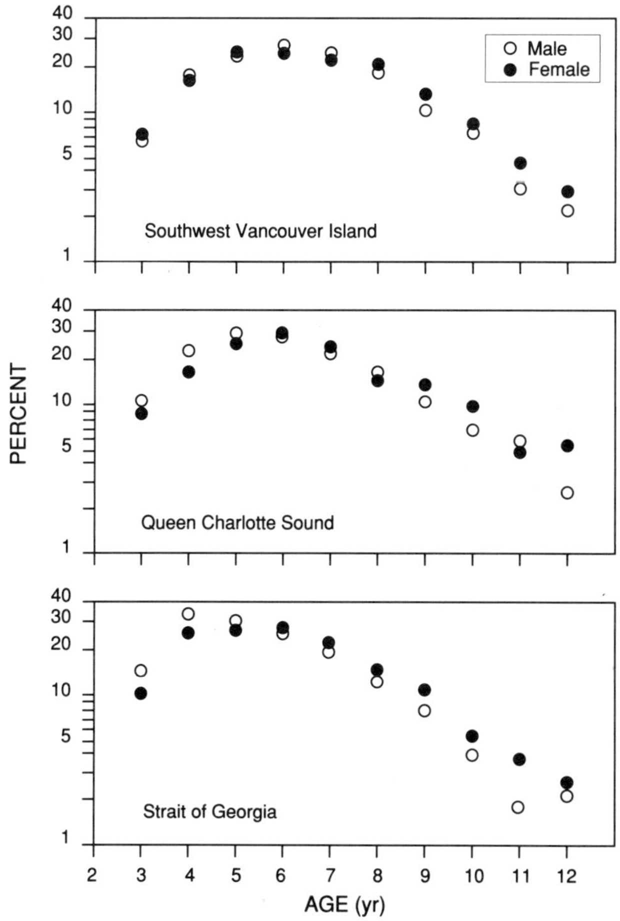

FIG. 43. Catch curves of lingcod showing the decline in the proportion at age of male and female lingcod in the commercial fishery off southwest Vancouver Island, in Queen Charlotte Sound, and in the Strait of Georgia.

In this model, we know the abundance of tagged lingcod in the first year (the number released). The CPUE in the first year is equal to q times the number of tagged fish released. The parameters to be estimated are σ and q. Survival from fishing mortality is estimated as $1 - qE$, where E is the average (1982–86) fishing effort. The harvest or exploitation rate is estimated as $\mu_t = qC_t/I_t$ where μ_t is the exploitation rate in year t.

Our estimate of annual survival from natural causes was 0.46. The 95% confidence interval ranged between 0.34–0.59. Our estimate of survival from fishing was 0.72. Our estimate of the overall survival is equal to the product of the two survival components or 0.33. Estimates of the exploitation rates for 1984 and 1985 were 0.33 and 0.52, respectively. Unfortunately, this model does not account for tag losses by fish after their release, non-reporting of tag recoveries by fishermen, or mortality due to tagging. It is important to note that these three sources of error result in tags being lost to the experiment and result in an underestimate of survival and exploitation rates.

In the Strait of Georgia there are no records that allow estimates of the catch per unit effort throughout the intensive fishery of the early to mid-1900s. It was pointed out earlier that there was a dramatic decline since the 1940s in both the commercial catch and the number of boats fishing lingcod in the Strait of Georgia. Beginning in 1967, we were able to estimate the CPUE for the lingcod handline fishery. We calculated the CPUE for two areas in the Strait of Georgia that were once sites of large commercial catches; the Campbell River area in the northern Strait of Georgia (Fig. 44) and the Gulf Island area in the southern Strait of Georgia (Fig. 45).

In the Campbell River area, CPUE remained fairly stable between 1967–76. Thereafter, CPUE declined steadily to a record low in 1986. The relatively high

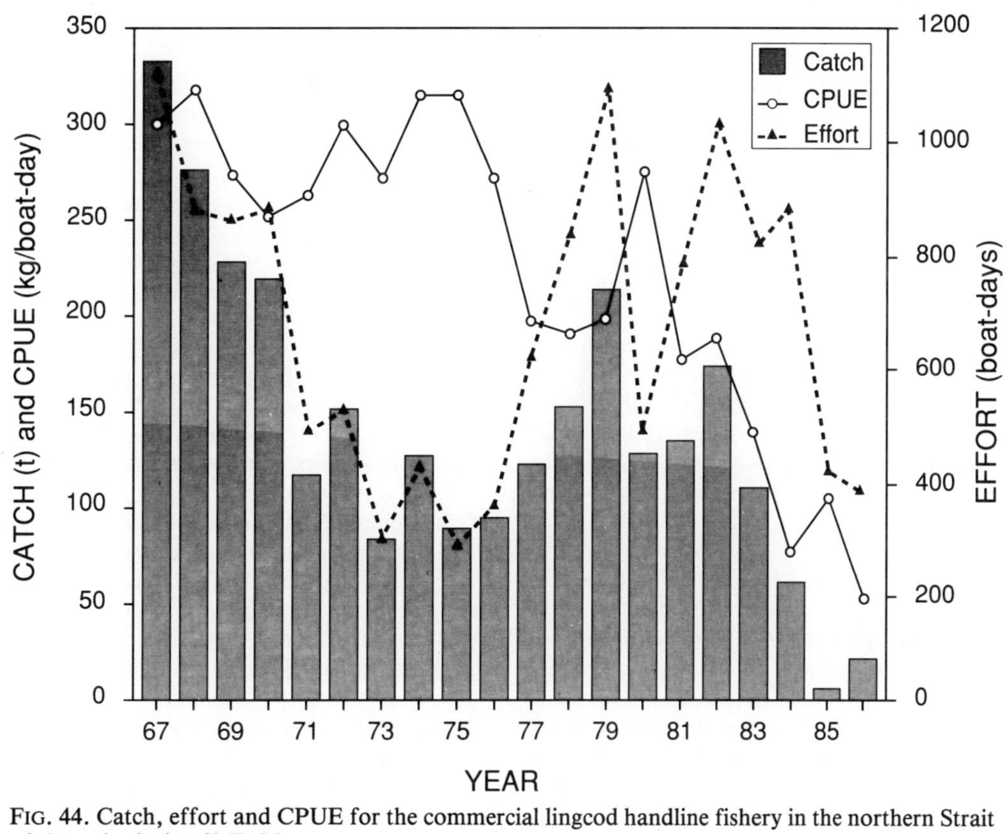

FIG. 44. Catch, effort and CPUE for the commercial lingcod handline fishery in the northern Strait of Georgia during 1967–86.

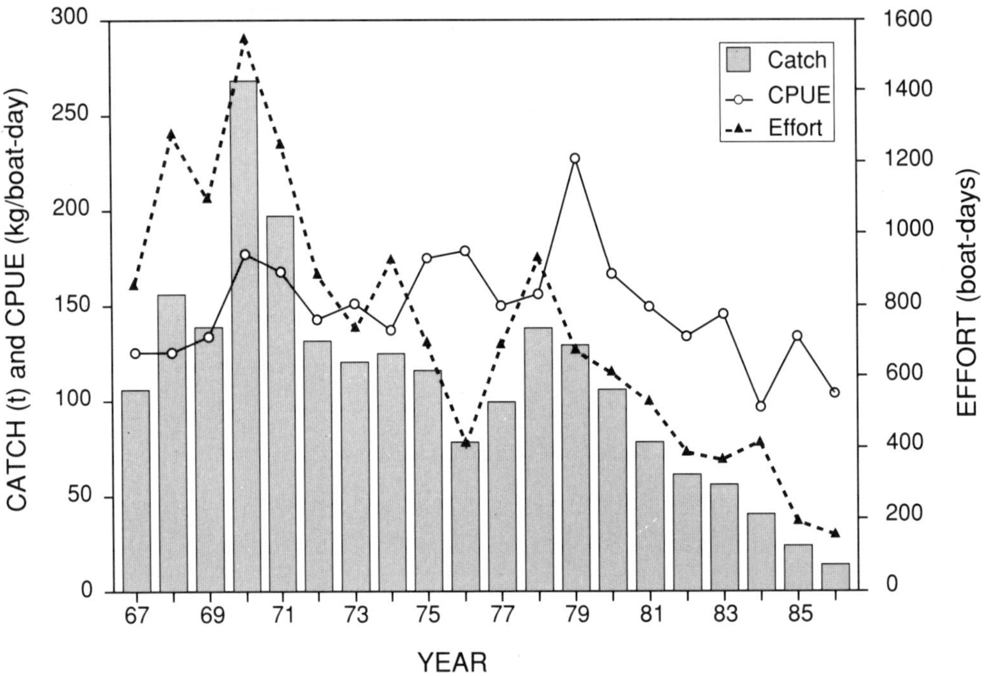

FIG. 45. Catch, effort and CPUE for the commercial lingcod handline fishery in the southern Strait of Georgia during 1967–86.

CPUEs between 1967–76 were associated with unusually low levels of fishing effort. However, the CPUE in recent years was low despite low levels of fishing effort. Similarly, in the Gulf Island area, the CPUE remained fairly stable from 1967–78. Catch, effort and CPUE have declined since that time. These findings indicate the decline of this once important fishery has continued into the 1980s.

Catches have also declined in the recreational line fishery. Recreational line catches averaged 80 000 lingcod/yr (125 t) between 1980 and 1986. The sports catch has decreased continuously since 1984. What is important to note is that together with catches by scuba divers of about 50 000 lingcod/yr (80 t), the sports catch now exceeds the catch in the commercial fishery.

Furthermore, a large portion of the sports catch consists of immature fish, and this overfishing of juveniles results in a reduction of the number of fish that reach maturity. This could ultimately result in reduced egg deposition and because lingcod stocks are localized, the potential for reduced egg production is a serious problem.

The theory that lingcod populations consist of local stocks that can easily be overfished was first expressed in a similar way more than 50 years ago. For example, in an article published in the *Canadian Fisherman* in 1926 (before the intensive fishery of the 1930s and 1940s), G. V. Wilby stated: "Naturally those areas which are closest to the market have been fished first, and it seems that some of these areas are now devoid of lingcod through intense fishing." Toivo Laine, who fished lingcod in the Strait of Georgia from 1944 to 1985 observed that reefs that became overfished during his early years as a fisherman never regained their original level of productivity even after the dramatic decline in fishing effort after the 1940s. Also, the Matsunaga family of Campbell River have witnessed similar declines, and the virtual collapse of the fishery in the Campbell River area from Chatham Point to Shelter Point.

The long-term decline in catch and fishing effort from an estimated 200 vessels in the 1940s to less than 10 bona fide fishermen in the 1970s and 1980s clearly reflects the decline in abundance. It is obvious that lingcod in the Strait of Georgia simply could not sustain the intense fishery that existed. Today, we believe the current population is still being overfished despite our attempts to regulate the Strait of Georgia fishery.

Some regulation of this fishery has been implemented. In fact, the first regulatory measures to protect lingcod stocks were implemented in the early 1920s. This was a winter fishing closure designed to protect spawning stocks in the Strait of Georgia. Initially, the fishery was closed between January and February in the Gulf Island region. After studying the spawning habits it was realized that the closure was not sufficient to protect spawning stocks. This in turn resulted in a fishing closure from November 15–April 15 for all waters off eastern Vancouver Island, including the Strait of Georgia and Juan de Fuca Strait. However, as yet there have been no legal restrictions on the catch of lingcod in other areas, except for a coastwide minimum size limit in commercial catches of 58 cm in length, implemented originally as a weight limit in 1942.

If the present trend of declining abundance continues, it may be necessary to consider other management strategies to protect lingcod in this area. These could take the form of area closures in those areas where stocks continue to decline, as well as the imposition of a size limit on lingcod caught in the sports fishery to ensure adequate survival to maturity. Even more drastic measures were taken in parts of Puget Sound, where a total fishing closure was imposed in 1978 by the Washington State Department of Fisheries. This closure followed a long-term decline in catch beginning in the late 1940s.

In the offshore areas the catch is almost entirely taken by commercial vessels and the data on the size of catches appear reliable (Fig. 46–48). We used the time series of catch and fishing effort along with information on growth and recruitment to estimate survival from natural and fishing mortality. Estimates of lingcod growth were calculated from tagging data. Growth was measured from the relationship between fish size at the time of tagging and their size one year later. Estimates of recruitment were made from the time series (1956–86) of size frequency data.

This model states that the abundance in year t is equal to the abundance that survived fishing mortality in year t-1 plus the increase in population from new recruits and growth times survival from natural mortality.

The estimate of survival from natural mortality, σ, was 0.76. The overall survival from both natural mortality and from fishing was essentially the same. In other words, the effect of the fishery could not be measured. The 95% confidence intervals for survival from natural mortality ranged between 0.65–0.97. An upper limit of the exploitation rate was 0.24 in 1984 and 0.39 in 1985. This corresponds to a lower limit on survival from fishing of 0.83.

Results from the fishery model differ significantly from the tagging model. The fishery data suggest that the present fishery is having little or no effect on the stock. The tagging data suggest that the exploitation rate is high. If the effects of tag loss, tagging mortality and non-reporting of tags were considered, the exploitation rate would be even higher.

There are a number of possible reasons why the estimates of survival based on the tagged population differ from estimates based on the untagged population. Tagged fish may behave differently from the untagged population, or there may actually be differences in exploitation rates of tagged and untagged fish. Also the tagging experi-

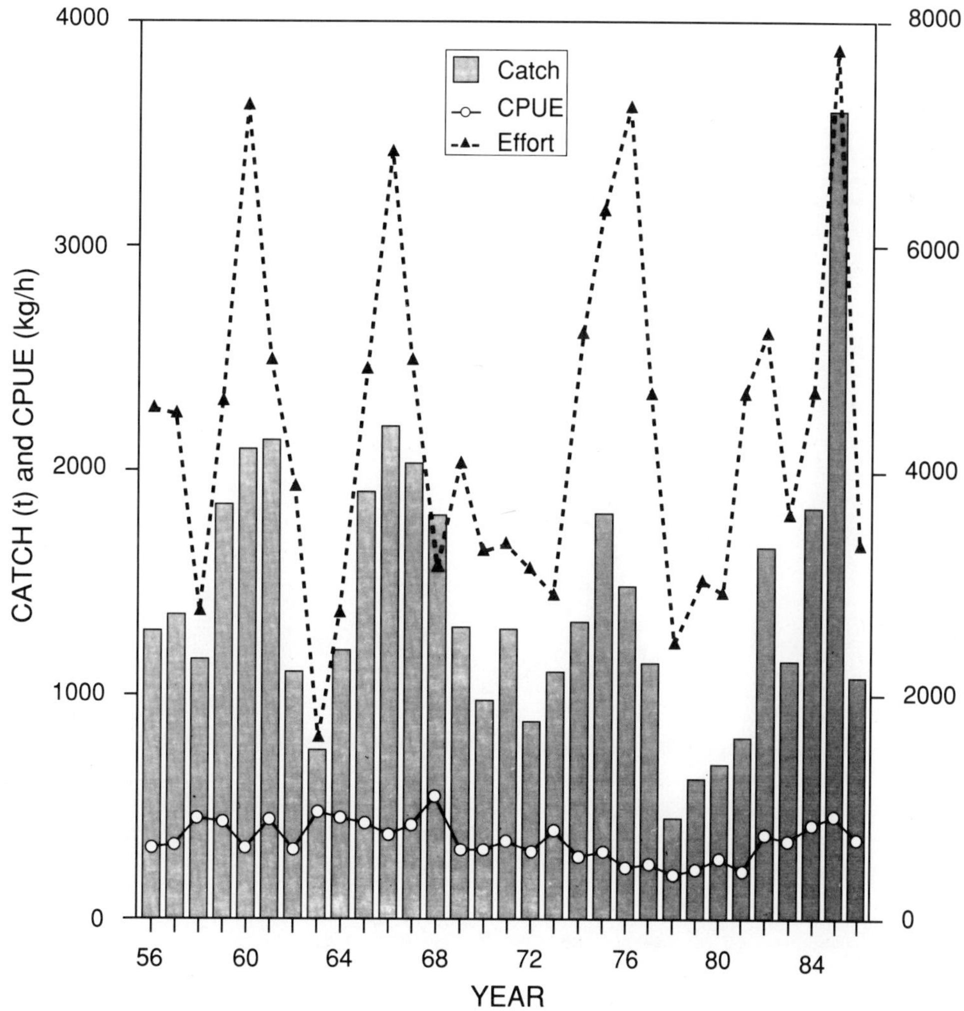

FIG. 46. Catch, effort and CPUE for the commercial lingcod trawl fishery off southwest Vancouver Island during 1956–86.

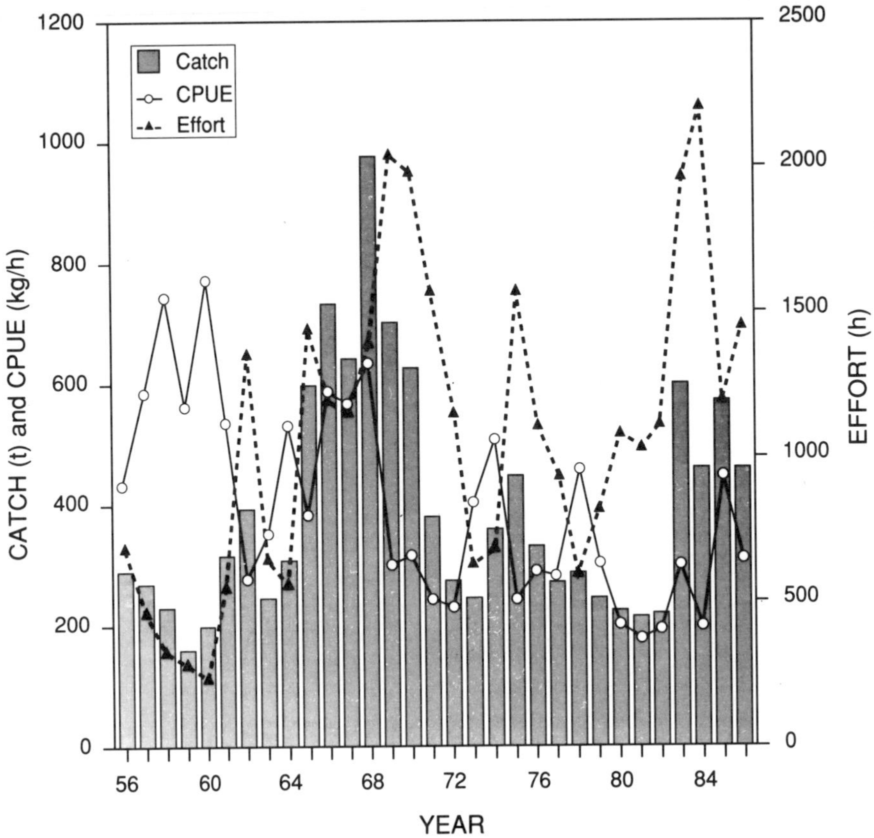

FIG. 47. Catch, effort and CPUE for the commercial lingcod trawl fishery off northwest Vancouver Island during 1956–86.

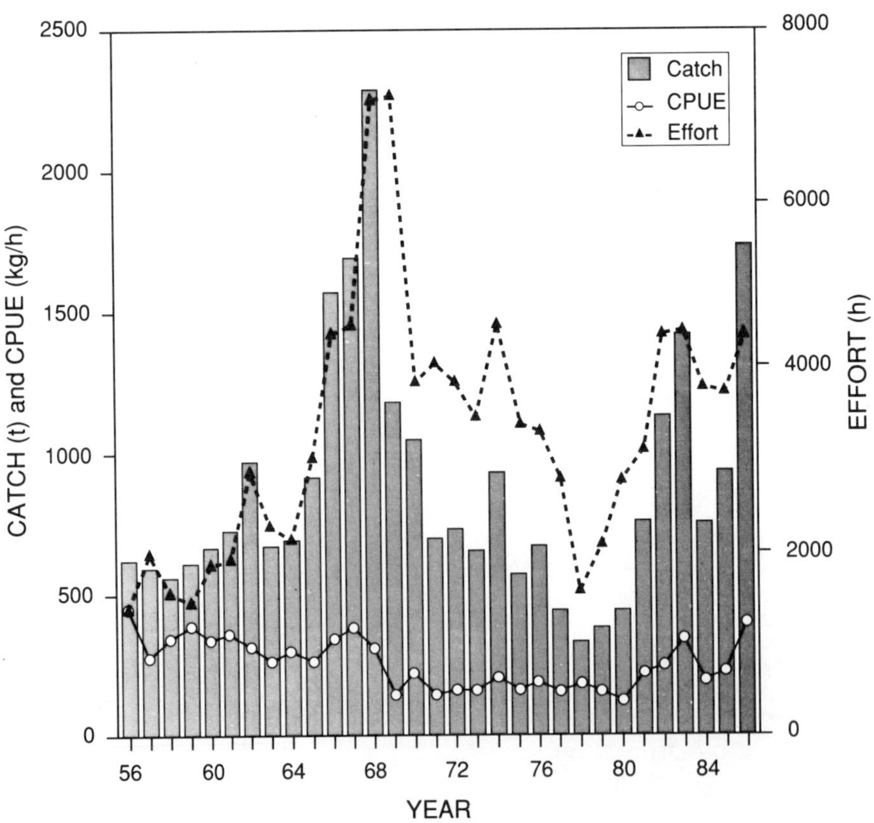

FIG. 48. Catch, effort and CPUE for the commercial lingcod trawl fishery in Queen Charlotte Sound during 1956–86.

ment took place on trawling grounds, and lingcod that reside in these areas may experience higher levels of fishing. Additionally, it was mentioned earlier that tag loss, mortality due to tagging, and non-reporting of tag recaptures can also affect survival and exploitation rates estimated from tagging experiments. It seems clear, therefore, that further studies are needed to test whether the differences between the two models are real.

Furthermore, the maximum catch that the stock off southwest Vancouver Island can sustain without causing long-term depletion of the stock is not known. We do know that the average catch of 1400 t/yr appears to be having little effect on stock size.

Lingcod in other offshore areas have been subjected to similar levels of fishing. However, there are fewer data for these areas and only the fishery model could be applied. Based on fishery data, the 95% confidence interval for survival from natural mortality off northwest Vancouver Island ranged between 0.67–0.88. The extreme value for natural survival corresponds to survival from fishing of 0.9. The upper limit of exploitation was 0.42 in 1984 and 0.23 in 1985. The upper limit of the exploitation rate was 0.19 in 1984 and 0.22 in 1985. The estimate of natural survival in Queen Charlotte Sound ranged between 0.78–0.99 with corresponding estimates of survival from fishing of 0–0.15. As found for stocks off Vancouver Island, the population of lingcod off Vancouver Island is sustainable at historical catch levels.

Despite strong fluctuations in recruitment to offshore stocks, there is some evidence that recruitment has declined over the last 10–15 years. Although we do not know the long term sustainable catches for lingcod stocks off Canada's west coast, recruitment should be monitored as an indicator of future stock sizes.

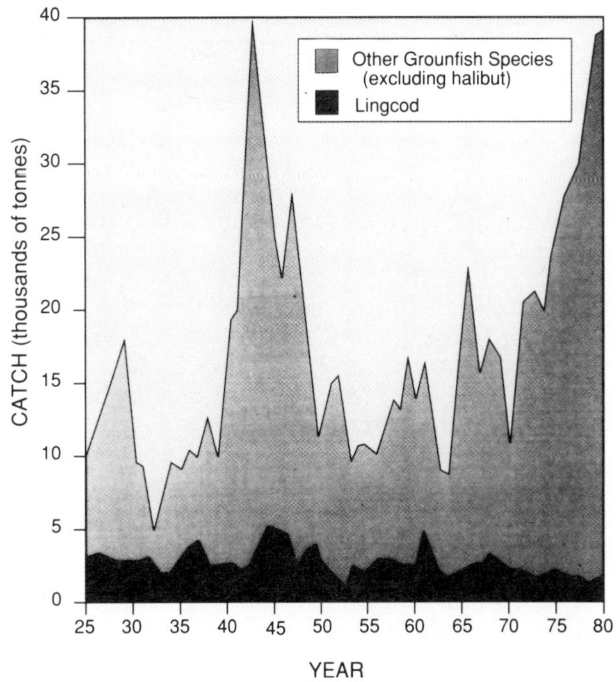

FIG. 49. Commercial catch of lingcod and other groundfish species, excluding halibut, caught off British Columbia by Canadian fishermen, 1930–80.

Markets and Economic Value

The landed value of the Canadian commercial fishery averaged nearly $1 million/yr during 1970–80. In 1985 the landed value increased to $3.4 million, the highest year on record. The high catches in 1985 were in part due to better markets and suggest the demand for lingcod may be expanding.

The commercial fishery supplies two somewhat different and competing markets. On the one hand, the handline fishery supplies a fresh product sold mainly as fillets and steaks to local restaurants and retail outlets. A portion of the catch is also exported to other Canadian and U.S. markets. In contrast, the more productive bottom trawl fishery services a broader based market that is less dependent on fresh fish markets. It nonetheless includes lingcod sold locally as well as those exported. During periods of high lingcod production, the prices paid to fishermen often decline. Nevertheless, over the long-term, the demand for lingcod has remained constant (Fig. 49).

The economic value of the recreational fisheries is difficult to measure. The total annual value of the recreational fishery in British Columbia has been estimated at approximately $30 million in 1981. This is based on the value generated from fishing anglers and not on the market value of the fish. This estimate does not include expenditures on boats, fishing equipment or the value of the scuba fishery. Fishing effort directed at non-salmonids accounts for about 40% of the total sports fishery of which lingcod and rockfish are the main contributors. Using this estimate, it is evident that the recreational fishery for lingcod is significant.

Summary

Our account of the biology and fishery of lingcod was written so that anyone interested in the marine resource can appreciate the importance of conserving this magnificent fish. We described the importance of the species to the first Canadians and later to the settlers coming from the east. There was no threat to the survival of stocks in these early times but it was only a matter of time before the balance of nature shifted in favour of man. Today, some stocks are in danger. In fact, some stocks urgently require protection. Overfishing seems to be an inevitable consequence of our expanding population. Preventing overfishing may not be possible because of the limitations of our science; recognizing that it has occurred and restoring stocks is something we can do.

When you read this summary it is important that you remember that much of the knowledge of lingcod biology was obtained in recent times and there still is much to learn. It is also important to recognize that fisheries science is quite young and again there is much to learn. Finally, we want to remind the reader that we omitted some technical descriptions so that our account would be more readable to a general audience. However, we did not exclude all technical information to show the kind of information that is needed and how it is used to manage fisheries.

Lingcod are distributed in coastal marine waters off the west coast of North America. In Canadian waters, they occur in the Strait of Georgia, off the west coast of Vancouver Island and in Queen Charlotte Sound. Adults are commonly found in rocky areas at depths of 10–100 m, where currents are strong.

Adult lingcod move very little. There are, however, localized seasonal movements related to maturation and spawning. Juveniles disperse over a wider geographic range.

Lingcod are demersal spawners. Eggs are deposited as a mass which is firmly attached to a rocky substrate. Spawning in Canadian waters begins in early December and continues through March. Prior to spawning, maturing males and females segregate. Males begin to aggregate in rocky, current-swept areas during November–December at depths of 5–60 m. At this time males become territorial and select nesting sites. Maturing females aggregate at deeper depths and remain segregated from males except for a brief spawning period.

The mean age and size at 50% maturity for females has ranged from 3 to 5 yr and from 61 to 76 cm. Most females are mature by age 7 and 80 cm. Most males are mature at age 2 and 50 cm. It appears that larger females spawn first, but it is not known if larger males move into the spawning areas first and select the most suitable sites.

The most suitable areas for egg deposition occur in rocky crevices in areas of strong currents. Spawning appears to occur at night. After spawning, males remain close to the nest and protect the egg masses from predators. The presence of the males is essential for successful hatching, which takes an average of 7 weeks.

Larval lingcod enter surface waters during March–May following hatching and become dispersed by surface currents. In the Strait of Georgia, they are more abundant in nearshore areas than in more exposed open waters. At this time, the larvae are 6–10 mm in length. By early April the majority of lingcod are about 20 mm long and by late April they are about 30 mm long. By late May–early June, these lingcod are about 70–80 mm in length and resemble adult lingcod in colouration and external morphology. In June, juvenile lingcod move from the surface to the bottom.

Juvenile age-0 lingcod first reside in shallow water near kelp and eelgrass beds. During the summer, they become dispersed from these relatively shallow nearshore areas to a wide range of depths. Age-1 lingcod are found in nearshore areas, segregated from older age-classes. At age-2 they begin to move into habitats typical of adult lingcod.

During larval development, diet consists mainly of copepods. As the larvae grow, larger food items, particularly larval herring, are ingested. Just prior to the transition to bottom life, young lingcod begin to prey on newly metamorphosed juvenile herring. It appears that this change in diet is associated with the transition to bottom life. During the juvenile period, lingcod continue to feed on herring as their major food. Adult lingcod are opportunistic carnivores and eat a variety of invertebrates and fish species including smaller lingcod.

Lingcod growth during the first few years is rapid. Young-of-the-year lingcod average 21 cm in length late in the first year. Age-1 lingcod average 33 cm by the second year and age-2 fish average 45 cm late in the third year. After age 2, females grow faster than males. At age 3 females are 3 cm larger and by age 13 they are 20 cm larger. The difference in size between males and females is particularly noticable in weight. By age 13 the difference in weight is an average of 12 kg or 26 lb. The rate of growth varies substantially between northern and southern areas off the British Columbia coast. This indicates that lingcod in these areas are different stocks. There is very little variation in the mean length at age within stocks.

Lingcod first become vulnerable to fishing at age 2. They do not contribute significantly to the commercial fishery until age 4 and are not fully recruited until age 6. Although a number of age-classes contribute to the commercial fisheries, strong year-classes make a major contribution to the catch. In recent years, offshore catches have been dominated by the 1977–78 and the 1980–82 year-classes.

Lingcod are fished commercially mainly by handline and trawl gear. The handline fishery is conducted in the Strait of Georgia and was once responsible for most of the commercial landings from Canadian waters. Since the 1940s, handline landings have declined steadily from an average of about 3 400 t/yr during the 1930s and 1940s to less than 1 000 t/yr during the 1970s. Catches during 1980–85 averaged 277 t/yr. The collapse of the commercial handline fishery probably was a result of overfishing. Beginning in the 1940s, the rapidly developing trawl fishery took a greater share of the annual landings of lingcod. The trawl fishery extended the range of exploitation outside the Strait of Georgia and is most active off the west coast of Vancouver Island and in Queen Charlotte Sound. Canadian trawl landings increased from 600 t in 1945 to 2 900 t in 1968. From 1969–80, domestic trawl landings averaged 1 400 t/yr. During 1980–85, landings increased to an average of 2 800 t/yr. Trawl catches in 1985 of 4 853 t were the highest on record.

Lingcod are also caught by sports fishermen and scuba divers. The size of the catch from the recreational fishery was first estimated in 1980. During 1980–85, recreational line catches accounted for an estimated 35% of the number of lingcod landed from the Strait of Georgia. During this period, an estimated average of 80 000 lingcod/yr were landed by sports line fishermen. A recent survey of scuba divers indicates lingcod is the principal fish species sought by spear fishermen and that catches are significant. Rough calculations indicate catches of 50 000 lingcod may have been taken in 1983.

Stocks off the west coast of Vancouver Island and in Queen Charlotte Sound have undergone marked fluctuations in abundance since 1956. During 1982–85 annual catches have been increasing as a result of strong recruitment. There is no evidence of long-term overfishing in these areas.

The economic value of the domestic commercial fishery, based on landed value, averaged nearly $1 million/yr during 1970–80 but rose to more than $3 million in 1985. The economic value of the sports fishery for lingcod cannot be accurately determined, but it is evident that it is an important fishery.

The size of any population is dependent on the number of eggs produced and larval survival. Survival is affected by oceanographic conditions and the abundance of predators and prey species. Many fishes release eggs directly into the ocean where they are dispersed over great distances prior to hatching. Unlike these fishes, lingcod concentrate their eggs in one mass. The larvae school close to shore and dispersion is restricted to local areas. The movement from the surface to the bottom within a few months of hatching also reduces the amount of dispersion. Because adult lingcod move very little once they become established in an area, there is only limited opportunity for lingcod to disperse from the spawning grounds. The restricted dispersal of young and the reduced movement of adults mean that there is very limited recruitment from other neighbouring stocks and that it is important to protect local stocks from overfishing.

While there are aspects that remain to be studied, our current understanding of the life history of lingcod indicates that management strategies must ensure that individual stocks are not overfished. Although the current strategy of winter fishing closures protects lingcod during the spawning and nesting period, stocks in parts of the Strait of Georgia are overfished and require additional protection such as size limits, area closures and, if necessary, total closures.

Lingcod have been an important part of the west coast heritage. They provided a readily accessible food source for native Indians and early settlers. They provided a valuable commercial and sports fishery in the Strait of Georgia and now off the west coast of Vancouver Island. However, lingcod are vulnerable to modern fishing methods and have been overfished in some areas. It is essential that lingcod fishing be controlled in the future, if our children and their children are to enjoy these colourful and impressive animals in their natural environment.

Source Material

ALVERSON, D. L. 1960. A study of animal and seasonal bathymetric catch patterns for commercially important groundfishes of the Pacific northwest coast of North America. Pac. Mar. Fish. Comm. Bull. 4.

BAGENAL, T. B. 1973. Fish fecundity and its relationship with stock and recruitment. Rapp. R. -V. Cons. Int. Explor. Mer 164: 186–198.

BARGMANN, G. G. 1982. The biology and fisheries for lingcod (*Ophiodon elongatus*) in Puget Sound. Wash. Dept. Fish. Tech. Rep. 66: 69 p.

BARNETT, G.H. 1955. The Coast Salish of British Columbia. The University Press, University of Oregon, Eugene, OR, USA.

BEAMISH, R. J. AND D. CHILTON. 1977. Age determination of lingcod (*Opiodon elongatus*) using dorsal fin-rays and scales. J. Fish. Res. Board Can. 34: 1305–1313.

BEAMISH, R. J., AND G. A. MCFARLANE. 1983. Use of oxytetracycline and other methods to validate a method of age determination for sablefish. *In* Proc. Int. Sablefish Symp., Lowell Wakefield Fisheries, Symposia Series, University of Alaska. Alaska Sea Grant Rep. 83-8: 95–116.

BEAMISH, R. J., K. R. WEIR, J. R. SCARSBROOK, AND M. S. SMITH. 1976. Growth of young Pacific hake, walleye pollock, Pacific cod and lingcod in Stuart Channel in 1975. Fish. Res. Board Can. MS Rep. 1399: 25 p.

1978. Growth of young Pacific hake, walleye pollock, Pacific cod and lingcod in Stuart Channel, British Columbia in 1976. Fish. Mar. Serv. MS Rep. 1518: 111 p.

BERNICK, K. 1983. A site catchment analysis of the Little Qualicum River site: A wet site on the east coast of Vancouver Island. Ottawa: National Museum of Man Mercury Series, Archaeological Survey of Canada, Paper 118.

BEVERTON, R. J. H., AND S. H. HOLT. 1957. On the dynamics of exploited fish populations. Fish. Invest. Ser. II, XIX: 533 p.

BLAXTER, J. H. S. 1969. Development: eggs and larvae, p. 177-252. *In* W. S. Hoar and D. J. Randall, [ed.] Fish physiology, Vol. III. Academic Press, New York, N.Y.

BLAXTER, J. H. S., AND G. HEMPEL. 1963. The influence of egg size on herring larvae (*Clupea harengus* L.). J. Cons. Cons. Int. Explor. Mer 28: 211–240.

BORDEN, C. E. 1975. Origins and development of early northwest coast culture to about 3000 B.C. Ottawa: National Museum of Man Mercury Series, Archaeological Survey of Canada, Paper 45.

CARLSON, C. 1979. The early component of Bear Cove. Can. J. Archaeol. 3: 177–193.

CASS, A. J., AND R. J. BEAMISH. 1983. First evidence of validity of the fin-ray method of age determination for marine fishes. N. Am. J. Fish Manage. 3: 182–188.

CASS, A. J., R. J. BEAMISH, AND M. S. SMITH. 1984. Study of the biology of lingcod off the west coast of Vancouver Island, M/V *Arctic Harvester*, November 22–December 2, 1977. Can. Data Rep. Fish. Aquat. Sci. 461: 71 p.

CASS, A. J., E. CAMERON, AND I. BARBER. 1983. Lingcod tagging study off southwest Vancouver Island, M/V *Pacific Eagle*, July 14–27, 1982. Can. Data Rep. Fish. Aquat. Sci. 406: 84 p.

CASS, A. J., G. A. MCFARLANE, K. RUTHERFORD, AND J. BARBER. 1984. Lingcod tagging study in the Strait of Georgia, November 1982–March 1983. Can. MS Rep. Fish. Aquat. Sci. 1791: 49 p.

CASS, A.J., G.A. MCFARLANE, M. S. SMITH, I. BARBER, AND K. RUTHERFORD. 1986. Lingcod tagging in the Strait of Georgia, 1983–84. Can. MS Rep. Fish. Aquat. Sci. 1875: 49 p.

CASS, A. J., L. J. RICHARDS, AND J. T. SCHNUTE. 1988. A summary of fishery and biological data for lingcod off the southwest coast of Vancouver Island. Can. Tech. Rep. Fish. Aquat. Sci. 1618: 33 p.

CASS, A.J., J.R. SELSBY, AND L.J. RICHARDS. 1986. Lingcod maturity cruise off southwest Vancouver Island and in Queen Charlotte Sound, R/V *G.B. Reed,* August 20–September 5, 1985. Can. Data. Rep. Fish. Aquat. Sci. 594: 35 p.

CASS, A. J., AND J. R. SCARSBROOK. 1984. A preliminary study of variability in year-class abundance of post-larval and juvenile lingcod in the Strait of Georgia during 1980–82. Can. MS Rep. Fish. Aquat. Sci. 1755: 27 p.

CASS, A. J., M. S. SMITH, I. BARBER, AND K. RINHOFER. 1983. A summary of lingcod tagging studies conducted in 1978 by the Pacific Biological Station. Can. Data Rep. Fish. Aquat. Sci. 417: 283 p.

CHATWIN, B. M. 1954. Growth of young lingcod. Fish. Res. Board Can. Pac. Prog. Rep. 99: 14–17.

1956. Age and growth of lingcod (*Ophiodon elongatus*). Fish. Res. Board Can. Pac. Prog. Rep. 105: 22–26.

1958. Mortality rates and estimates of theoretical yield in relation to minimum commercial size of lingcod (*Ophiodon elongatus*) from the Strait of Georgia, British Columbia. J. Fish. Res. Board Can. 15: 831–849.

CHILTON, D. E., AND R. J. BEAMISH. 1982. Age determination methods for fishes studied by the Groundfish Program at the Pacific Biological Station. Can. Spec. Publ. Fish. Aquat. Sci. 60: 102 p.

CUSHING, D. H. 1967. The grouping of herring populations. J. Mar. Biol. Assoc. U.K. 47: 193–208.

DEMPSTER, R. P., AND W. F. ROHRS. 1983. Spawning record of the lingcod, *Ophiodon elongatus* (Girard, 1854), in captivity. J. Aquacult. Aquat. Sci. III: 41–45.

DERISO, R. B. 1980. Harvesting strategies and parameter estimation for an age-structured model. Can. J. Fish. Aquat. Sci. 37: 268–282.

DURBIN, A. G. 1979. Food selection by plankton feeding fishes, p. 203–218. *In* Predation-prey systems in fisheries management. Sports Fishing Institute, Washington, DC.

FLADMARK, K. R. 1982. An introduction to the prehistory of British Columbia. Can. J. Archaeol. 6: 96–156.

FORRESTER, C. R. 1969. Life history information on some groundfish species. Fish. Res. Board Can. Tech. Rep. 105: 17 p.

1973. The lingcod (*Ophiodon elongatus*) in waters off western Canada. Fish. Res. Board Can. MS Rep. 1266: 27 p.

GIORGI, A. E. 1981. The environmental biology of the embryos, egg masses and nesting sites of the lingcod, *Ophiodon elongatus*. Nat. Mar. Fish. Serv. NWAFC Processed Rep. 81-06: 107 p.

GROSSE, D. J. 1982. An experiment in the artificial rearing of lingcod (*Ophiodon elongatus* Girard) for purposes of enhancement. MS Thesis, University of Washington, Seattle, WA.

HAIST, V., M. STOCKER, AND J. F. SCHWEIGERT. 1985. Stock assessments for British Columbia herring in 1984 and forecasts of the potential catch in 1985. Can. Tech. Rep. Fish. Aquat. Sci. 1365: 53 p.

HART, J. L. 1943. Migration of lingcod. Fish. Res. Board Can. Pac. Prog. Rep. 57: 3–7.

1967. Fecundity of lingcod. J. Fish. Res. Board Can. 24: 2495–2489.

HEMPEL, G. 1965. On the importance of larval survival for the population dynamics of marine food fish. Rep. Calif. Coop. Oceanic. Fish. Invest. 10: 13–23.

JEWELL, E. D. 1968. Scuba diving observations on lingcod spawning at a Seattle breakwater. Wash. Dep. Fish. Res. Pap. 3: 27–36.

JONES, R. 1973. Density-dependent regulation of the numbers of cod and haddock. Rapp. P.-V. Reun. Cons. Int. Explor. Mer 164: 156–173.

JORDON, D. S., AND C. H. GILBERT. 1882. Notes on fishes of the Pacific coast of the U.S. Rapp. P.-V. Reun. Cons. Int. Explor. Mer 4: 29–70.

KENDALL, A. W., AND B. VINTER. 1984. Development of hexagrammids (Pisces: Scorpaeniformes) in the northeastern Pacific Ocean. Nat. Mar. Fish. Serv. NOAA Tech. Rep. NMFS 2: 44 p.

KETCHEN, K. S. 1976. Catch and effort statistics of the Canadian and United States trawl fisheries in waters adjacent to the British Columbia coast. 1950–1975. Fish. Mar. Serv. Data Rep. 6: 578 p.

[ed.]. 1980. Assessment of groundfish stocks off the west coast of Canada (1979). Can. Data Rep. Fish. Aquat. Sci. 185: 213 p.

LARIVIERE, M. G., D. D. JESSUP, AND S. B. MATHEWS. 1981. Lingcod (*Ophiodon elongatus*), spawning and nesting in San Juan Channel, Washington. Calif. Fish. Game 67: 231–239.

LORD, J. K. 1866. The naturalist in Vancouver Island and British Columbia 1: 358 p. Richard Bentley, London.

LOW, C. J., AND R. J. BEAMISH. 1978. A study of the nesting behavior of lingcod (*Ophiodon elongatus*) in the Strait of Georgia, British Columbia. Fish. Mar. Serv. Tech. Rep. 843: 27 p.

MACGREGOR, M. 1982. The tidal sportfishing diary program report on the pilot years 1979–80. Dep. Fish. Oceans, Vancouver, B.C.

MANN, R. H. K., AND C. A. MILLS. 1979. Demographic aspects of fish fecundity. Symp. Zool. Soc. Lond. 44: 161–177.

MCELDERRY, H. I., AND L. J. RICHARDS. 1984. Recreational scuba diving in the Strait of Georgia. An analysis of the distribution of diving effort and the importance of collecting marine animals. Can. MS Rep. Fish. Aquat. Sci. 1794: 61 p.

MILLER, D. J., AND J. J. GEIBEL. 1973. Summary of blue rockfish and lingcod life histories; a reef ecology study; and giant kelp, *Macrocystis pyrifera*, experiments in Monterey Bay, California. Calif. Dep. Fish. Game 158: 137 p.

MOULTON, L. L. 1977. An ecological analysis of fishes inhabiting the rocky nearshore regions of northern Puget Sound, Washington. Ph.D. dissertation, University of Washington, Seattle, WA.

PEARSE, P. H. 1982. Turning the tide. A new policy for Canada's Pacific Fisheries. The Commission on Pacific Fisheries policy, final report. Vancouver, B.C.

PHILLIPS, J. B. 1959. A review of the lingcod, *Ophiodon elongatus*. Calif. Fish. Game 45: 19–28.

PHILLIPS, A. C., AND W. E. BARRACLOUGH. 1977. On the early life history of the lingcod (*Ophiodon elongatus*). Fish. Mar. Serv. Tech. Rep. 756: 35 p.

RASS, T. S. [ED.]. 1962. Greenlings: taxonomy, biology, interoceanic transplantation. Academy of Sciences of the USSR. Transactions of the Institute of Oceanology. Vol. 59. Translated from Russian. U.S. Department of Commerce, Springfield, VA, USA.

REEVES, J. E. 1966. An estimate of survival, mortality, and of the number of lingcod (*Ophiodon elongatus*) off the southwest coast of Vancouver Island, British Columbia. Wash. Dep. Fish. Research Paper 12: 55–66.

ROBINSON, C. K., L. A. LAPI, AND E. W. CARTER. 1982. Data collected during spiny dogfish (*Squalus acanthias*) stomach content survey near the Qualicum and Fraser rivers, April–May 1980–81. Can. Data Rep. Fish. Aquat. Sci. 325: 64 p.

ROEDEL, P. M. 1948. Common marine fishes of California. Calif. Dep. Fish. Game Fish Bull. 66: 150 p.

SCHNUTE, J. T., L. J. RICHARDS, AND A. J. CASS. 1989. Fish growth: investigations based on a size structured model. Can. J. Fish. Aquat. Sci. 46: 730–742.

1989. Fish survival and recruitment: investigation based on a size structured model. Can. J. Fish. Aquat. Sci. 46: 743–769.

SPALDING, D. J. 1964. Comparative feeding habits of the fur seal, sea lion and harbour seal on the British Columbia coast. Fish. Res. Board Can. Bull. 146: 52 p.

STOCKER, M. [ED.]. 1981. Groundfish stock assessments off the west coast of Canada in 1981 and recommended total allowable catches for 1982. Can. MS Rep. Fish. Aquat. Sci. 1626: 282 p.

THOMSON, J. A., AND A. N. YATES. 1961. British Columbia landings of trawl-caught groundfish by month, by minor statistical area. Fish. Res. Board Can. Circ. (statistical series) No. 4.

1961. British Columbia landings of trawl-caught groundfish by month, by minor statistical area. Fish. Res. Board Can. Circ. (statistical series) No. 5.

1961. British Columbia landings of trawl-caught groundfish by month, by minor statistical area. Fish. Res. Board Can. Circ. (statistical series) No. 2.

WEIR, K. R., R. J. BEAMISH, M. S. SMITH, AND J. R. SCARSBROOK. 1978. Hake and pollock study, Strait of Georgia bottom trawl cruise *G.B. Reed*, February 25–March 13, 1975. Fish. Mar. Serv. Data Rep. 71: 153 p.

WESTRHEIM, S. J. [ED.]. 1980. Assessment of groundfish stocks off the west coast of Canada in 1979 and recommended total allowable landings for 1980. Can. Data Rep. Fish. Aquat. Sci. 208: 265 p.

1983. A new method for allotting fishing effort to individual species in a mixed-species trawl fishery. Can. J. Fish. Aquat. Sci. 40: 352–360.

WESTRHEIM, S. J., AND R. P. FOUCHER. 1985. Relative fishing power for Canadian trawlers landing Pacific cod (*Gadus macrocephalus*) and important shelf cohabitants from major offshore areas of western Canada, 1960–81. Can. J. Fish. Aquat. Sci. 42: 1614–1626.

WILBY, G. V. 1937. The lingcod, *Ophiodon elongatus* Girard. Bull. Fish. Res. Board Can. 54: 24 p.

Glossary of Terms

Adult/juvenile
Terms which distinguish sexually mature fish (adults) from sexually immature fish (juveniles).

Age at maturity
The age at which an organism or group of organisms (e.g., year-class) spawns for the first time. The age at maturity for a year-class is often expressed as the age when 50% of the year-class spawns for the first time. It should be noted that the first spawning of a year-class may extend over several years.

Age determination
The process of measuring the age of a fish.

Age group (class)
All individuals of a given age within a population.

Age-specific mortality
Mortality expressed as a function of age.

Bottom trawl fishery
Fisheries conducted using a towed net from a vessel wherein the groundline of the net is in contact with the seabed during the fishing operation.

Catchability
The fraction of a fish stock which is caught by a unit of fishing effort.

Catch curve
A graph of the logarithm of the number of fish taken at successive ages.

Catch-per-unit-effort (CPUE)
The weight or number of organisms which are caught by a given unit of fishing gear, divided by the fishing effort applied to obtain that quantity of organisms.

Colonization
The settlement of a species or group of species in close association in a particular location.

Confidence interval
An interval which has a specified probability of containing a given parameter (i.e., mortality rate). A 95% level of statistical significance indicates the parameter must be accepted 95 times in 100.

Dorsal
The back of the fish.

Effort (i.e. hours of trawling)
Refers to the time spent when the fishing gear is effectively fishing. In the case of a bottom trawl, which is fished on the bottom, then the effort is the time elapsed between when the net arrives on the bottom until it leaves. In this report, fishing effort was standardized to account for variation in vessel horsepower among vessels engaged in the lingcod fishery.

Egg viability
The ability of an egg to develop normally.

Exploitation rate
The ratio of the catch in a given year to the population at the beginning of that year.

Fecundity
The reproductive capacity of an individual organism expressed as the number of eggs produced.

Fin ray
The bony structure comprising the fins of fishes. Cross-sections of fin rays are used to measure the age of some fishes.

Fishing mortality (rate)
Mortality in a stock arising from fishing operations.

Handline fishery
Commercial fisheries conducted using hook and line gear wherein a single fishing line is suspended per fisherman just off the bottom generally in rocky areas.

Hydrolyze
To undergo a chemical reaction with water.

Landed value
The dock-side value of commercially caught fish paid by the buyer to the fisherman.

Lateral line
A line along the sides of most fishes, often distinguished by differently coloured scales, which mark the lateral line sensory organ.

Length frequency distribution
The distribution of fish lengths in a sample.

Longline fishery
Commercial fisheries wherein fish are caught on hooks attached by short lines (ganglions) to a long mainline.

Model
A theoretical (mathematical) representation that obeys certain specified conditions used to estimate population parameters (i.e., fishing mortality) and gain an understanding of the population to which it is analogous in some way.

Natural mortality
Mortality in a stock arising from causes other than fishing including such causes as disease, senescence, predation, starvation, lethal environmental conditions and cannibalism.

Non-reporting of tags
Refers to the proportion of tagged fish caught after their release but not reported to those responsible for the tagging experiment.

Offshore fishery
In this report the offshore fishery refers to all fisheries occurring outside the inside waters of eastern Vancouver Island.

Ova
A term for unfertilized eggs.

Overfishing
A term describing a situation wherein the yield from a fishery is below the level of some optimal yield per recruit directly as a result of catching too many fish.

Population
Any self-perpetuating group of organisms within which there is complete gene flow.

Population growth
The rate of change in population or stock size per unit time measured in terms of its biomass or number of fish as opposed to individual growth.

Recruitment
The process whereby fish are introduced or are "recruited" to populations, stocks or fishing grounds. This process, while generally referring to the process wherein young fish enter a fishable stock, may also describe the process wherein older fish may immigrate to a stock from other stocks or areas.

Reproductive potential
Total potential offspring from an organism or group of organisms (year-class).

Seine (purse seine) fishery
Commercial fisheries wherein a body of fish is surrounded by a net which is then gathered or "pursed" at the bottom.

Size limit
Refers to a minimum or maximum fish size restriction, usually expressed as fish length, imposed for stock conservation reasons through fisheries regulation.

Stock
The part of a population, which may include the entire population, that is under consideration with regard to exploitation or potential exploitation.

Strong year-class
A year-class of much greater abundance than is usual, such that its presence tends to distort or alter the long-term age or size structure of a population during its presence in that population.

Sunken gill-net
Fisheries conducted by anchoring a net along the seabed. This method is of historical significance only and is no longer allowed in Canadian waters.

Survival rate
Number of fish alive after a specified time interval, divided by the initial number.

Sustainable yield
The quantity which a fishery will yield indefinitely, under some specific exploitation pattern.

Tagging experiment
A study that involves marking and releasing a subset of a population that along with their recovery provides insight into the unmarked population. Tagging experiments are used to estimate migration rates and to estimate growth and survival rates.

Tagging mortality
Mortality associated with tagging operations. This may be divided into two segments: (1) the initial traumatic mortality caused by the capture of the organisms and by tagging procedures; and (2) the longer-term mortality caused by the presence of the tag on or in the organism and the effects on its normal activity.

Tag loss
Pertains to the proportion of fish in a tagging experiment that inadvertently lose or shed their tags. Fish are often tagged with two tags to estimate the rate of tag loss. The number of double-tagged fish recaptured with only one tag is compared to the number recaptured with two tags.

Time series
Refers to data collected routinely at specific time intervals (e.g., annually).

Trap/pot fishery
The method of fishing employing baited traps.

Troll fishery
Commercial and recreational fisheries wherein a lure or bait is towed at the end of fishing lines throughout the water column.

Ventral
The side opposite the back.

Vulnerability
The proportion of individuals in a stock or year-class susceptible to capture by fishing gear.

Year-class
All individuals of a species which are spawned or born in a given year (cohort). Commonly used in the context of a population or stock.

Yield per recruit
The expected yield in weight from a single recruit.

Appendix

TABLE 1. Commercial lingcod catch (tonnes) in British Columbia.

	Canada			Canada–U.S.	
Year	Line	Trawl	Total	Trawl	Total
1889	169	—	169	—	—
1890	127	—	127	—	—
1891	93	—	93	—	—
1892	109	—	109	—	—
1893	291	—	291	—	—
1894	201	—	201	—	—
1895	178	—	178	—	—
1896	181	—	181	—	—
1897	181	—	181	—	—
1898	329	—	329	—	—
1899	339	—	339	—	—
1900	346	—	346	—	—
1901	310	—	310	—	—
1902	340	—	340	—	—
1903	386	—	386	—	—
1904	459	—	459	—	—
1905	421	—	421	—	—
1906	448	—	448	—	—
1907	470	—	470	—	—
1908	397	—	397	—	—
1909	313	—	313	—	—
1910	373	—	373	—	—
1911	848	—	848	—	—
1912	883	—	883	—	—
1913	917	—	917	—	—
1914	1346	—	1346	—	—
1915	1368	—	1368	—	—
1916	1274	—	1274	—	—
1917	1379	—	1379	—	—
1918	2994	—	2994	—	—
1919	3198	—	3198	—	—
1920	2313	—	2313	—	—
1921	1973	—	1973	—	—
1922	1905	—	1905	—	—
1923	1973	—	1973	—	—
1924	2790	—	2790	—	—
1925	2177	—	2177	—	—
1926	2654	—	2654	—	—
1927	3147	—	3147	—	—
1928	3201	—	3201	—	—
1929	3057	—	3057	—	—
1930	3064	—	3064	—	—
1931	3215	—	3215	—	—
1932	2519	—	2519	—	—
1933	2540	—	2540	—	—
1934	3014	—	3014	—	—
1935	3962	—	3962	—	—
1936	4346	—	4346	—	—
1937	2702	—	2702	—	—
1938	2933	—	2933	—	—
1939	2995	—	2995	—	—
1940	3002	—	3002	—	—
1941	2577	—	2577	—	—
1942	2680	—	2680	—	—
1943	3700	—	3700	—	—
1944	5312	—	5312	—	—
1945	4033	630	4663	—	—
1946	3407	659	4066	—	—
1947	2200	243	2443	—	—
1948	3701	451	4152	—	—
1949	3842	737	4579	—	—
1950	2137	787	2924	—	—
1951	2142	850	2992	—	—
1952	1981	507	2677	—	—
1953	1376	370	1746	—	—
1954	1875	597	2472	1248	3123
1955	1528	777	2305	1924	3452
1956	1855	1116	2971	2060	3915
1957	2007	985	2992	1964	3971
1958	1748	961	2709	1787	3535
1959	1496	1132	2628	2856	4352
1960	1660	1078	2738	2926	4586
1961	1450	1313	2763	3080	4530
1962	1763	952	2715	2303	4066
1963	1414	650	2064	1553	2966
1964	1311	1282	2593	2348	3659
1965	1731	1742	3473	3564	5294
1966	1213	2043	3256	4466	5678
1967	1325	1961	3286	4275	5600
1968	1130	2971	4101	5265	6395
1969	1316	1820	3136	3114	4431
1970	1497	1440	2937	2395	3892
1971	1176	1555	2731	2207	3383
1972	1250	1038	2288	1538	2788
1973	862	1205	2067	1766	2628
1974	943	1507	2450	2311	3254
1975	890	1895	2785	2680	3570
1976	863	1368	2231	2195	3059
1977	971	1176	2147	1528	2499
1978	646	909	1555	934	1580
1979	703	1164	1867	1184	1887
1980	598	1352	1950	1353	1951
1981	592	1758	2350	1758	2350
1982	960	2888	3848	2888	3848
1983	790	2995	3785	2995	3785
1984	726	2973	3699	2973	3699
1985	817	4853	5670	4853	5670

TABLE 2. Summary of lingcod tagging studies.

Country and year	Tagging location (reference)	No. tagged	No. recovered	Comments
Canada				
1939–43	Strait of Georgia (Hart 1943; Chatwin 1956)	1993	336	
Canada				
1944–45 1950 1953–54	Strait of Georgia (Chatwin 1956)	889	214	
1947, 1949 1950–51	West coast Vancouver Island (Chatwin 1956)	263	77	
USA				
1960	West coast Vancouver Island (Reeves 1966)	437	284	
Canada				
1964	West coast Vancouver Island (Forrester 1973)	2000	774	
USA				
1963–65	California	19	2	
1967–71	(Miller and Geibel 1973)	238	15	
USA				
1969–71	Oregon (Demory, Oregon Dep. Fish. Game, pers. comm.)	19	17	
USA				
1977–80	Northern Puget Sound (La Riviere et al. 1981)	1536	87	
USA				
1978–81	Northern Puget Sound (Bargmann 1982)	1275	73	
Canada				
1978	Strait of Georgia (Cass et al. 1983)	752	7	Results as of Dec 31, 1988
Canada				
1982–85	Strait of Georgia (Cass et al. 1984, 1986)	10 293	882	Studies in progress; results as of Dec 31, 1988
Canada				
1978	West coast Vancouver Island (Cass et al. 1983)	2997	48	Results as of Dec 31, 1988
Canada				
1982	West coast Vancouver Island (Cass et al. 1983)	7429	2607	Studies in progress; results as of Dec 31, 1988

TABLE 3. Recoveries of tagged lingcod by year and distance travelled from releases off the southwest coast of Vancouver Island in 1982.

Distance (km)	Recovery year					
	1982	1983	1984	1985	1986	Total
< = 10	1356	601	225	126	11	2319
11–50	37	56	13	16	6	128
>50	3	14	6	4	2	29
Total	1396	671	244	146	19	2476

TABLE 4. Recoveries of tagged lingcod by year of recovery and distance travelled from tag releases in the Strait of Georgia in 1982–83, 1983–84 and 1984–85.

Distance (km)	Recovery year				
	1st	2nd	3rd	4th	Total
< = 10	577	99	13	0	689
11–50	26	4	0	1	31
>50	2	0	0	0	2
Total	605	103	13	1	722

TABLE 5. Summary of lingcod spawning and nest guarding observations.

	Wilby (1937)	Jewell (1968)	Low and Beamish (1978)	La Riviere et al. (1981)	Giorgi (1981)
Spawning depth	3–9 m below low tide.	2–11 m of the plus 2 tide level.	To 30 m from high tide level.	2–20 m below mean low low water.	10–22 m below mean low low water. Shallowest spawning occurred in slow current areas.
Spawning period	Mid-December to March. Peak activity in January.	Mid-December to mid-March. Peak activity mid-January to mid-February.	Early January–mid march. Peak activity late January to early February.	Mid-December–early April. Peak activity late February–early March.	Not examined.
Spawning location	Crevices in boulder area, little marine growth.	Crevices or rock ledges.	Crevices or rock ledges.	Crevices or under boulders.	Noted spawning on cobble-oyster shell substrate in slow current areas. Suggests that water movement is an important stimulus for nest site selection.
Spawning process	Female deposits successive layers of ova to be fertilized by male. Male remains and guards nests.	Male may fertilize ova from more than one female.	Male may fertilize ova from more than one female.	Male may fertilize ova from more than one female.	Not examined.
Guarding behaviour	Male shapes egg mass and fans eggs with pectoral fins for aeration, aggressively repels all intruders including invertebrates.	Male guards one or more nests. No fanning of eggs. Not aggressive. Subtidal nests guarded more closely than intertidal nests. Other males will guard a nest if original male removed.	Male guards one or more nests. No fanning. Not aggressive except to larger fish. Invertebrate and small fish predation ignored. Unguarded nests may be guarded by other males some of whom may not be mature.	Male guards one or more nests. No fanning. Non-aggressive and aggressive behaviour noted. Aggressive behaviour in larger males only. Invertebrate and small fish predation ignored, aggressive to larger fish. Males will guard two nests in close proximity.	Invertebrate predation ignored.

TABLE 5. (Continued) Summary of lingcod spawning and nest guarding observations.

	Wilby (1937)	Jewell (1968)	Low and Beamish (1978)	La Riviere et al. (1981)	Giorgi (1981)
Predators	Inference that male is able to repel all predators.	Predation important but except for some invertebrates, it was inferred that predation was highest in unattended nests.	Predation very important. Large number of fish and invertebrate predators present.	Impact of predation unknown. Predation by invertebrates was common. Vertebrate predation uncommon.	The extent of predation varied among areas. The regions of slow current velocity was where predators were in low abundance. Invertebrates were the most effective predators. The most common predators were gastropods but they consumed relatively few eggs.
Incubation period	About 6 wk. Male leaves at time of hatching.	Seven weeks, some nests still present as late as March 14.	More than 5 wk, estimated 7–8 wk but one nest 11 wk. Some nests still present in early June.	About 5–7 wk. Most nests hatched by mid-April. Viable nests may be present in June.	Not examined.
Hatching	No information.	Hatching started about mid-January with peak hatching mid-February–mid-March. Nests require 3–7 days to hatch. Males left area by mid-March — more than 90% of nests hatched.	Hatching started mid-March. Most of nest hatched in one day. Males left nests that hatched early in season shortly after hatching. Some males left later hatching nests before hatching. Egg viability reduced for nests hatching later in season. Ten to 27% of nests hatched. The rest were lost to predation.	Hatching started in February. Males remained in their territories until the nest hatched and occasionally after. Estimated hatching success of 40%. Losses due to predation or dislodgement could not be identified.	In laboratory studies most nests in high current velocities hatched within a few hours and completely hatched within 24–48 h. Hatching of nests from low current regions was more protracted. In the laboratory these nests hatched sequentially from the periphery inwards over a 2-week period. Mean egg survival in low current areas was 61%. The principal source of mortality was attributed to hypoxia.

TABLE 6. Fish and invertebrate predators of lingcod nests observed during scuba diving and hook and line surveys by area and study.

Common name	Scientific name	Area of observation	References
Kelp greenling	*Hexagrammos decagrammus*	Strait of Georgia	Low and Beamish (1978)
Striped seaperch	*Embiotoca lateralis*	Strait of Georgia	Low and Beamish (1978)
Longfin sculpin	*Jordania zonope*	Strait of Georgia Puget Sound	Low and Beamish (1978) Jewell (1968)
Puget Sound sculpin	*Artedius meanyi*	Strait of Georgia	Low and Beamish (1978)
Pile perch	*Rhacochilus vacca*	Puget Sound	Jewell (1968)
Scalyhead sculpin	*Artedius harringtoni*	Puget Sound	Moulton (1977) La Riviere et al. (1981)
Puget Sound rockfish	*Sebastes emphaeus*	Puget Sound	Giorgi (1981)
Yelloweye rockfish	*Sebastes ruberrimus*	Strait of Georgia	Cass, unpublished data
Quillback rockfish	*Sebastes maliger*	Strait of Georgia	Cass, unpublished data
Copper rockfish	*Sebastes caurinus*	Strait of Georgia	Cass, unpublished data
Cabezon	*Scorpaenichthys marmoratus*	Strait of Georgia Puget Sound	Cass, unpublished data Jewell (1968)
Red sea urchin	*Strongylo-centrotus franciscanus*	Strait of Georgia Puget Sound	Cass, unpublished data La Riviere et al. (1981) Moulton (1977) Giorgi (1981)
Green sea urchin	*Strongylo-centrotus drobachiensis*	Puget Sound	Moulton (1977) Jewell (1968)
Purple sea urchin	*Strongylo-centrotus purpuratus*	Puget Sound	La Riviere et al. (1981)
Sunflower star	*Pycnopoda helianthoides*	Strait of Georgia	Low and Beamish (1978)
Anemone	*Tealia crassicornis*	Strait of Georgia	Low and Beamish (1978)
Gastropod	*Calliostoma ligatum*	Puget Sound west coast Vancouver Island	Moulton (1977) La Riviere et al. (1981) Present study
Gastropod	*Amphissa columbiana*	Puget Sound west coast Vancouver Island	Moulton (1977) La Riviere et al. (1981) Present study
Starfish	*Pisaster ochraceus*	Puget Sound	Giorgi (1981)
Crab	*Phyllo- lithodes papillosus*	Puget Sound	Moulton (1977)
Sharp-nose crab	*Scyra acuthifrons*	Puget Sound	Moulton (1977)

TABLE 7. Mean length and estimated mean weight of lingcod off the west coast of Vancouver Island, 1977–85.

	Length (cm)		Weight (kg)		Weight (lb)	
Age	Male	Female	Male	Female	Male	Female
2	45	45	0.7	0.7	1.5	1.5
3	58	61	1.9	2.2	4.2	4.9
4	63	68	2.4	3.1	5.3	6.9
5	67	73	3.0	3.9	6.5	8.6
6	70	78	3.4	4.8	7.5	10.7
7	73	82	3.9	5.7	8.6	12.5
8	76	86	4.4	6.6	9.8	14.6
9	78	90	4.8	7.7	10.7	17.0
10	80	94	5.2	8.9	11.6	19.5
11	82	97	5.7	9.8	12.5	21.6
12	83	100	5.9	10.8	13.0	23.9
13	84	103	6.1	11.9	13.6	26.3
14	—	105	—	12.7	—	28.0
15	—	108	—	13.9	—	30.6
16	—	110	—	14.8	—	32.5
17	—	112	—	15.6	—	34.5
18	—	114	—	16.6	—	36.5
19	—	116	—	17.5	—	38.6
20	—	117	—	18.0	—	39.7